T0195519

Assessing the Assignment Policy for Army Women

Margaret C. Harrell · Laura Werber Castaneda · Peter Schirmer
Bryan W. Hallmark · Jennifer Kavanagh · Daniel Gershwin · Paul Steinberg

Prepared for the Office of the Secretary of Defense

Approved for public release; distribution unlimited

NATIONAL DEFENSE RESEARCH INSTITUTE

The research described in this report was prepared for the Office of the Secretary of Defense (OSD). The research was conducted in the RAND National Defense Research Institute, a federally funded research and development center sponsored by the OSD, the Joint Staff, the Unified Combatant Commands, the Department of the Navy, the Marine Corps, the defense agencies, and the defense Intelligence Community under Contract W74V8H-06-C-0002.

Library of Congress Cataloging-in-Publication Data is available for this publication.

ISBN 978-0-8330-4150-0

Photo courtesy of U.S. Army
Photographed by Air Force Staff Sgt. Cindy Haught
Photo is of Sgt. Katherine Tripp, from the 982nd Signal Company, as she prepares for a
mission in a Stryker vehicle departing from Forward Operating Base, Marez, Iraq.

The RAND Corporation is a nonprofit research organization providing objective analysis and effective solutions that address the challenges facing the public and private sectors around the world. RAND's publications do not necessarily reflect the opinions of its research clients and sponsors.

RAND® is a registered trademark.

© Copyright 2007 RAND Corporation

All rights reserved. No part of this book may be reproduced in any form by any electronic or mechanical means (including photocopying, recording, or information storage and retrieval) without permission in writing from RAND.

Published 2007 by the RAND Corporation
1776 Main Street, P.O. Box 2138, Santa Monica, CA 90407-2138
1200 South Hayes Street, Arlington, VA 22202-5050
4570 Fifth Avenue, Suite 600, Pittsburgh, PA 15213-2665
RAND URL: http://www.rand.org/
To order RAND documents or to obtain additional information, contact
Distribution Services: Telephone: (310) 451-7002;
Fax: (310) 451-6915; Email: order@rand.org

Preface

The current U.S. Department of Defense (DoD) policy for assigning military women dates to a 1994 memorandum from then–Secretary of Defense Les Aspin. During the ensuing years, the U.S. military has undergone significant technological and organizational transformation, which has resulted in changes in how the military organizes and fights. Specifically, the Army's recent transformation to modular brigades, as well as the differences between military missions in Iraq, and the global war on terrorism (GWOT) more generally, and military missions fought on linear battlefields during past military engagements, prompted concern among some members of Congress about the role of women in military operations in Iraq. Reflecting that, Section 541(b) of Public Law 109-163 requires the Secretary of Defense to submit a report on the current and future implementation of DoD policy for assigning military women.[1]

This monograph is intended as input in DoD decisionmaking and focuses on Army operations in Iraq. In particular, it focuses on the Army's brigade combat teams (BCTs) that deployed to Iraq in a modular configuration, paying specific attention to the new organic relationships of these BCTs with brigade support battalions (BSBs).

This research was sponsored by the Under Secretary of Defense for Personnel and Readiness and conducted within the Forces and Resources Policy Center of the RAND National Defense Research Institute, a federally funded research and development center spon-

---

[1] Public Law 109-163, National Defense Authorization Act for Fiscal Year 2006, January 6, 2006.

sored by the Office of the Secretary of Defense, the Joint Staff, the Unified Combatant Commands, the Department of the Navy, the Marine Corps, the defense agencies, and the defense Intelligence Community. The principal investigator is Margaret C. Harrell. Comments are welcome and may be addressed to Margaret_Harrell@rand.org.

For more information on RAND's Forces and Resources Policy Center, contact the Director, James Hosek. He can be reached by email at James_Hosek@rand.org; by phone at 310-393-0411, extension 7183; or by mail at the RAND Corporation, 1776 Main Street, Santa Monica, California 90407-2138. More information about RAND is available at www.rand.org.

Contents

Tables

Summary

Introduction

In January 1994, informed by the report of the Presidential Commission on the Assignment of Women to the Armed Forces, then–Secretary of Defense Les Aspin established the current DoD assignment policy for women in the military with a memorandum that stated that personnel can

> be assigned to all positions for which they are qualified, except that women shall be excluded from assignment to units below the brigade level whose primary mission is to engage in direct combat on the ground. . . .[1]

The same memorandum also promulgated a definition of direct combat on the ground:

> Direct ground combat is engaging the enemy on the ground with individual or crew served weapons, while being exposed to hostile fire and to a high probability of direct physical contact with the hostile force's personnel. Direct ground combat takes place well forward on the battlefield while locating and closing with the enemy to defeat them by fire, maneuver, or shock effect.

---

[1] Les Aspin, Secretary of Defense, "Direct Ground Combat Definition and Assignment Rule," memorandum, January 13, 1994. The memorandum is included as Appendix A.

The Aspin memorandum also indicated that the military services' policies and regulations could include certain restrictions on the assignment of military women

> where units and positions are doctrinally required to physically collocate and remain with direct ground combat units that are closed to women.

The Army policy for assigning women, Army Regulation (AR) 600-13,[2] predates the Aspin memorandum and is similar to, but not the same as, the DoD policy for assigning military women. AR 600-13 states,

> The Army's assignment policy for female soldiers allows women to serve in any officer or enlisted specialty or position except in those specialties, positions, or units (battalion size or smaller) which are assigned a routine mission to engage in direct combat, or which collocate routinely with units assigned a direct combat mission.[3]

Important to understanding the Army policy is recognizing that it defines direct combat differently from the DoD policy. The Army policy defines direct combat as follows:

> Engaging an enemy with individual or crew served weapons while being exposed to direct enemy fire, a high probability of direct physical contact with the enemy's personnel and a substantial risk of capture. Direct combat takes place while closing with the enemy by fire, maneuver, and shock effect in order to destroy or capture the enemy, *or while repelling the enemy's assault by fire, close combat, or counterattack.*[4]

---

[2]   Headquarters, U.S. Department of the Army, Army Regulation 600-13, Army Policy for the Assignment of Female Soldiers, March 27, 1992.

[3]   Headquarters, U.S. Department of the Army (1992, p. 1).

[4]   Headquarters, U.S. Department of the Army (1992, p. 5). Emphasis added.

There are several important differences between the Army and DoD policies. First, the DoD policy restricts the assignment of women to units whose *primary* mission is direct ground combat, whereas the Army restricts assignment to units that have a *routine* mission of direct combat. Second, the Army also restricts assignment to units that collocate with direct combat units. Third, the Army and DoD policies define combat differently: The Army's definition of direct combat includes a requirement that there be a risk of capture, but also includes "repelling the enemy's assault." These differences are significant, and it is notable that the Army did not update its policy when Congress repealed the legal restrictions against women serving in combat aircraft positions and on combatant ships nor when Aspin revised the DoD policy in 1994.

Nonetheless, as a result of the DoD policy change, Army units and occupations were opened to women, resulting in positions for women within the headquarters of maneuver and support brigades, as well as in the headquarters of other units types, such as the special forces group.[5] As of the end of fiscal year 2006, the active-component Army includes more than 48,000 women, who have the opportunity to serve in 92.3 percent of Army occupations; 70.6 percent of Army positions are open to women.

Given the Army's recent modularization, as well as the differences between military missions in Iraq in the context of the GWOT and military missions fought on the linear battlefields of past military engagements, concerns have arisen among some members of Congress and other interested parties as to whether the roles of Army women in Iraq are consistent with existing policies; thus, Public Law 109-164 requires the Secretary of Defense to submit a report on the current and future implementation of DoD policy for assigning military women. The research presented here is provided as input to the Office of the Secretary of Defense (OSD).

---

[5]   The Army assignment policy, AR 600-13 (Headquarters, U.S. Department of the Army, 1992), predates the Aspin memorandum (Aspin, 1994). Positions were opened to Army women in 1994 as a result of Secretary Aspin's removal of the risk rule from the DoD assignment policy.

This research assesses the extent to which current policy pertaining to the assignment of military women is appropriate for and reflected in Army doctrine, transformation, and operations in Iraq. More specifically, this study focused on answering three main questions: (1) Is there a shared interpretation of the assignment policy for Army women? (2) Is the Army complying with the assignment policy? and (3) Are the language and concepts in the assignment policy appropriate to future Army operations, given what we know about operations in Iraq?

In analyzing current policy and the Army's compliance with it, it is important to underscore the purpose of that policy, which limits the units to which women can assigned but not the ways in which female service members can be tasked or utilized once in the theater of operations.

Is There a Shared Interpretation of the Assignment Policy for Army Women?

In answering the first question, we find that neither the Army nor the DoD assignment policies for military women are clearly understandable. Our interviews with senior personnel from the Army, OSD, and the Joint Staff (JS), as well as our sessions with personnel recently returned from Iraq, confirm that there is no shared interpretation of the meanings of many of the words used in the policy, including *enemy, forward* or *well forward*, and *collocation*. This is the result of the many policy changes that have occurred since 1994, including the Army's transformation, as well as the nature of warfare in Iraq.

Given the lack of a common understanding of the "letter" of the policy, this research sought to ascertain whether there was a shared interpretation of the "spirit" of the policy. Although senior Army, OSD, and JS personnel fairly consistently portrayed the objectives of an ideal assignment policy, we find from interviews with and public statements by members of Congress and interviews and meetings with congressional staff that there is not much agreement among members of Congress. Further, there was not a majority consensus among the senior DoD interviewees regarding the objectives reflected in the current assignment policy.

Is the Army Complying with the Assignment Policy?

To answer the second question, we consider whether Army women are assigned to units proscribed in the assignment policy, i.e., whether Army women are assigned to direct combat units below the brigade level (or battalion-size or smaller)[6] and whether Army women are assigned to units that collocate with direct combat units. This monograph finds that the Army is complying with the DoD assignment policy, although it may not be complying with the separate Army assignment policy.

Are Army Women Assigned to Direct Combat Units Below the Brigade Level?

To determine whether women are assigned to direct combat units, it is important to consider whether women are assigned to maneuver units whose established primary mission is direct combat and whether support units have adopted direct combat missions.

Women are not assigned to maneuver units below the brigade level, and this complies with DoD and Army assignment policy. However, we found that, under certain circumstances, support units to which women are assigned are in relationships with maneuver units that differ very slightly from the actuality of being assigned to those maneuver units, and that, in some circumstances, members of such a support unit have a closer relationship with the maneuver unit than with the unit in their assigned chain of command. Although these assignments meet the "letter" of the assignment policy, the assignments may involve activities or interactions that framers of the policy sought to rule out and that today's policymakers may or may not still want to preclude.

---

[6] The DoD policy specifies "below the brigade level" whereas the Army policy specifies "battalion size or smaller."

Support unit personnel in Iraq, including women, were trained, prepared, and expected to defend themselves and their fellow personnel. Level I self-defense (against snipers, agents, saboteurs, or terrorist activities) was a routine mission among support units in Iraq. Interviewed service members confirmed that women in support units were actively involved in routine self-defense missions that included providing security for their own units, providing personal security for leadership, and, in some cases, providing security for other support units. The Army assignment policy includes "repelling the enemy's assault by fire, close combat, or counterattack" in the definition of direct combat and states that units that have a *routine* mission of such direct combat should be closed to women.[7] However, it is unclear whether level I self-defense is included in "repelling the enemy's assault," as another interpretation of this phrase of the policy is that it intentionally refers to the documented core mission of maneuver units, which is to "close with the enemy by means of fire and maneuver to destroy or capture enemy forces, or to *repel their attacks by fire, close combat, and counterattack*."[8]

If individual or small-group self-defense is included in the direct combat definition, then assigning women to units that routinely conducted self-defense was not in compliance with the Army policy. Compliance with this interpretation would have significant consequences and could close to women many, if not all, support units. Of note, however, is that such assignments would not violate the DoD assignment policy, which does not include repelling the enemy's assault in its definition of direct combat and which also closes to women only those units whose *primary* mission is direct combat.

Are Army Women Assigned to Units That Are Collocated with Direct Combat Units?

The Army policy states that women cannot be assigned to units that routinely collocate with direct combat units. The interpretation of this restriction depends on the definition of *collocation*. The Army policy

---

[7] Headquarters, U.S. Department of the Army (1992, p. 5).

[8] Headquarters, U.S. Department of the Army, *The Brigade Combat Team*, FM 3-90.6, Washington, D.C., August 4, 2006b, p. 2-1. Emphasis added.

provides a definition that appears to be contingent on the proximity of units, but some interviewees maintained that collocation is defined as both unit interdependence and physical proximity. We consider the implications of both definitions in this monograph.

We find considerable evidence that support units are collocated with direct combat units if the definition is based purely on proximity. However, if the definition of collocation is based on interdependency *and* proximity, the evidence is inconclusive. Although some might maintain that the ability of maneuver units to accomplish their missions independently, even for a limited number of days, means that the support units and the maneuver units are not collocated, others might argue that the maneuver units' dependence on forward support companies (FSCs), or even the support units' dependence on maneuver units for security, *does* constitute collocation. Neither proximity nor the combination of proximity and interdependence of support and maneuver units would be inconsistent with the DoD assignment policy, which does not include a collocation restriction, but the assignment of women to support units in Iraq may not be consistent with the Army's assignment policy, given that support units are in proximity to maneuver units and that they may be both proximate to and interdependent with maneuver units.

It is important to note that this ambiguity stems from the fact that the policy as stated is ambiguous in its intent. Is the intent to keep women from being assigned to *any* units that might, for a variety of security or operational reasons, share the same base with maneuver units? If so, this is not being achieved in units to which women are assigned in Iraq. Or is the intent simply to keep women from areas that are especially vulnerable to large-scale enemy attack? If so, then policy is consistent with current operations in Iraq.

Are the Language and Concepts in the Assignment Policy Appropriate to Future Army operations?

Even though the DoD and Army assignment policies relate solely to assignment and do not have any bearing on the utilization of individual personnel in military operations, we nevertheless saw the need to consider the assignment policy in the context of Army operations in Iraq. Our view is that these operations provide insights into the actual roles and risks experienced by women under current policy and, perhaps, also offer information relevant to how a new assignment policy should be crafted. Additionally, the Army policy is written such that determining whether Army women are being assigned to appropriate Army units, based on some aspects of the Army policy, depends on the activities of the units while deployed.

Therefore, to help policymakers decide whether the policy should change, this study also evaluated whether the concepts and language in the current policy for assigning women are appropriate for future military operations, given the Army's experience in Iraq. As a point of departure in this evaluation, we include the attitudes and perceptions of returned service members regarding the current assignment policy. Not surprisingly, many personnel recently returned from Iraq did not know about the policy, as they were not generally involved in the assignment of personnel to units. Those who were familiar with the assignment policy did not generally find it understandable or useful. Some felt it was a backward step from operations in which women were involved that were being conducted successfully in Iraq. Some personnel also expressed the opinion that adherence to the policy, if it were interpreted strictly, would be a backward step in the successful execution of the mission in Iraq, in which women have been involved in many aspects of operations, and that a strict interpretation of the assignment policy could even prevent women from participating in Army operations in Iraq, which would preclude the Army from completing its mission. We acknowledge that many of their perceptions may be based on misinterpretations of the policy, but nonetheless, their attitudes confirm both the confusion about the actual meaning of the assignment policy as well as the necessity of an analysis of the appropriateness of the specific wording of the assignment policy.

While the perceptions of returned service members provide useful context and a valuable understanding of the conduct of Army operations in Iraq, this portion of our analysis focused primarily on the appropriateness of the specific wording of the policy. In many ways, the language and concepts in the current policy for assigning military women do not seem well suited to the type of operations taking place in Iraq. The focus on a defined enemy and the linear battlefield does fit the picture of traditional military operations but is inappropriate to Iraq. Further, the Army restriction on women in units that have a mission to repel an enemy's assault requires clarification with regard to the inclusion of self-defense missions. If it does include self-defense missions, that clause seems inappropriate to current operations in Iraq and potential future operations there and elsewhere, including those conducted during the period of insurgency and sectarian violence.

The appropriateness of other aspects of the assignment policy is a matter of interpretation and judgment. For example, the restriction against assigning women to maneuver units does keep women from being part of units that initiate direct combat or that close with the enemy. However, none of these restrictions preclude women from interacting closely with maneuver unit personnel or from interacting with the enemy or with potential enemy personnel. These restrictions do ensure that support units (and the women in them) are trained and mentored by other support unit personnel while in garrison, but they do not ensure closer proximity to the support unit in the chain of command than to maneuver units while in the theater. These restrictions could be interpreted to exclude support units from the benefit of extra security provided by maneuver units and could eliminate female service members from jobs they have performed successfully in Iraq. Indeed, a very strict interpretation of the Army's assignment policy could preclude some women from deploying to Iraq. Finally, granted that the assignment policy focuses on the assignment of women to units, judging the appropriateness of this focus requires consideration of the employment of women in the theater. Military effectiveness and flexibility entail adapting to changes in enemy strategy, tactics, and weapons, and this implies that commanders may need to employ military resources, including individual women and units with women, in ways not initially envisioned in policy and possibly not well addressed in doctrine.

The Iraq example has shown how the application of the current assignment policy has led to the employment of units including women in ways that are consistent with DoD policy but might not be consistent with the Army's assignment policy, and yet, based on our interviews and focus groups, has been consistent with maintaining unit effectiveness and capability.

Recommendations

This research effort set out to assess the extent to which the assignment policy for women is appropriate to, and reflected in, Army doctrine, transformation, and operations in Iraq. The intent is not to prescribe policy, but rather to report research findings about the assignment policy and Army operations in Iraq and to identify issues in policy, interpretation of policy, doctrine, or employment for DoD's consideration. The critical first issue is whether there should remain an assignment policy for military women. Removing the assignment policy for military women would be tantamount to asserting that women should be permitted to serve in combat units. This monograph is not intended to inform the policy debate that would emerge from completely removing the assignment policy.

If there continues to be an assignment policy for women in the military, then we recommend these considerations to guide its design, implementation, and any legal reporting requirement:

- Recraft the assignment policy for women to make it conform—*and clarify how it conforms*—to the nature of warfare today and in the future, and plan to review the policy periodically.
- Make clear the objectives and intent of any future policy.
- Clarify whether and how much the assignment policy should constrain military effectiveness and determine the extent to which military efficiency and expediency can overrule the assignment policy. For example, does the requirement to provide Congress with 30 days' notice of any change to the policy constrain military effectiveness, and would a longer time requirement do so?

- Consider whether a prospective policy should exclude women from units and positions in which they have successfully performed in Iraq.
- Given that the assignment policy is unusual because of the legal requirement to report policy changes to Congress, consider the extent to which an individual service policy should differ from overall DoD guidance. Recognize that those differing policies could present reporting challenges.
- Determine whether an assignment policy should restrict women from specified occupations or from both occupations and units. For example, should women be assigned to supply positions in an infantry battalion?
- Determine whether colocation (proximity) is objectionable and whether collocation (proximity and interdependence) is objectionable and clearly define those terms, should they be used in the policy.
- If unit sizes (or levels of command) are specified in the assignment policy, make apparent the reason and intent for specifying unit size, given that modularization, as well as the evolving battlefield, may blur or even negate this distinction.
- Consider whether the policy should remain focused on the assignment to units rather than the individual employment of women.

Acknowledgments

The authors thank the staff of the Office of the Deputy Under Secretary of Defense for Military Personnel Policy for their support, specifically, William Carr, Sheila Earle, COL Charles Armentrout, Stanley Cochran, Al Bruner, and especially Sam Retherford for his extensive coordination during this effort. The research team benefited from tremendous cooperation from the Office of the Deputy Chief of Staff for Personnel (Army G-1), from which we thank COL Dennis Dingle, LTC Lynn Jackson, LTC Kathryn Ensworth of the G-1 Human Resources Policy Directorate, and Kathy Dillaber. We benefited considerably from the expertise of COL Paul D. Thornton, Plans Division, Army G-1. We also appreciate the data support provided by Barbara Balison, Defense Manpower Data Center; CW4 Linda Johnson, Military Awards Branch, U.S. Army Human Resources Command; and Frederick Weatherson, Training and Analysis Division, U.S. Army Human Resources Command. In addition, we thank Craig Hayes, Fred Wham, and Brad Cox at the Center for Army Lessons Learned for sharing their observations and available supporting materials.

We appreciate the thoughtful input provided by the senior personnel interviewed for this research, including LTG James L. Campbell, David S. C. Chu, Kathryn A. Condon, RADM Donna Crisp, MG Rhett Hernandez, Mark R. Lewis, Mark D. Manning, Paul Mayberry, BG K. C. McClain, LTG Walter L. Sharp, and BG Robert H. Woods.

We could not have completed this work without the candid participation of many Army personnel in confidential interviews and discussion groups. We benefited from the tremendous assistance of the officers who coordinated our visit to the Combined Logistics Captains Career Course (CLC3) at Fort Lee, Virginia, and to a recently returned Army unit whose name and location are not disclosed for confidentiality reasons. We also thank the officers from the Army War College who were willing to speak to us informally.

Finally, we note the valuable contributions of RAND colleagues: Kevin Brancato for his help in assessing the women deployed to Iraq, Gary Cecchine for his expertise throughout the project and his participation in our fieldwork, Samantha Merck and Vicki Wunderle for their administrative support, and Lauren Skrabala for editing the monograph. LTC Robert Bradford, LTC Nanette Patton, and MAJ Nancy Blacker also provided important contributions to this effort during their tenure as RAND Arroyo Center military fellows. We appreciate the comments and careful reviews of RAND colleagues Lynn E. Davis, Lisa S. Meredith, Henry A. Leonard, and James Hosek. This monograph was also read and commented on by GEN John Keane, whose comments helped to clarify and strengthen the final document. While we appreciate the improvements provided by those who have reviewed and commented on this monograph, we emphasize that the findings and recommendations, as well as any errors, are those of the authors alone.

Abbreviations

AO	area of operations
AOC	area of concentration
AR	Army Regulation
ARS	armed reconnaissance squadron
BCT	brigade combat team
BSB	brigade support battalion
BSTB	brigade special troops battalion
CAB	Combat Action Badge
CBRN	chemical, biological, radiological, and nuclear
CLC3	Combined Logistics Captains Career Course
CMF	career management field
CSS	combat service support
DCPC	Direct Combat Probability Code
DoD	U.S. Department of Defense
eMILPO	Electronic Military Personnel Office
FOB	forward operating base
FSB	forward support battalion
FSC	forward support company

GWOT	global war on terrorism
HBCT	heavy brigade combat team
HHC	headquarters company
HUMINT	human intelligence
IBCT	infantry brigade combat team
IED	improvised explosive device
JS	Joint Staff
METL	mission-essential task list
MOS	military occupational specialty
MRE	meal, ready to eat
MSR	military supply route
NATO	North Atlantic Treaty Organization
NCO	noncommissioned officer
OPCON	operational control
OSD	Office of the Secretary of Defense
PST	private security team
PSYOP	psychological operations
ROE	rules of engagement
RPG	rocket-propelled grenade
SBCT	Stryker brigade combat team
SRC	standard requirements code
TTPs	tactics, techniques, and procedures
UAV	unmanned aerial vehicle
UIC	unit identification code
UCMJ	Uniform Code of Military Justice

CHAPTER ONE

Introduction

U.S. Department of Defense and Army Assignment Policies for Military Women

In January 1994, informed by the report of the Presidential Commission on the Assignment of Women to the Armed Forces, then–Secretary of Defense Les Aspin established the current U.S. Department of Defense (DoD) assignment policy for women in the military with a memorandum specifying rules to replace the prior "risk rule."[1] The risk rule had precluded women from serving in occupations or units characterized by the risk of exposure to direct combat, hostile fire, or capture. The current DoD assignment policy for military women instead establishes that military women can

> be assigned to all positions for which they are qualified, except that women shall be excluded from assignment to units below the brigade level whose primary mission is to engage in direct combat on the ground. . . .[2]

The same memorandum also promulgated a definition of direct combat on the ground:

---

[1] Aspin (1994). This memorandum is included as Appendix A. This action by Secretary Aspin followed the congressional repeal in 1993 of the laws that had precluded women from serving in combat aircraft positions or on combatant ships. Although combat aircraft and combatant ships had been closed to women by law, women's roles in ground units have always been constrained by policy rather than law.

[2] Aspin (1994).

Direct ground combat is engaging an enemy on the ground with individual or crew served weapons, while being exposed to hostile fire and to a high probability of direct physical contact with the hostile force's personnel. Direct ground combat takes place well forward on the battlefield while locating and closing with the enemy to defeat them by fire, maneuver, or shock effect.

The Aspin memorandum also indicated that the military services' policies and regulations could include certain restrictions on the assignment of military women:

where units and positions are doctrinally required to physically collocate and remain with direct ground combat units that are closed to women; where units are engaged in long range reconnaissance operations and Special Operations Forces missions; and where job related physical requirements would necessarily exclude the vast majority of women Service members.

The services may include these restrictions at their discretion. Such restrictions are permitted by DoD policy but they are not constraints of that policy.

The Army policy for assigning women, Army Regulation (AR) 600-13,[3] predates the Aspin memorandum and is similar to, but not the same as, the DoD policy for assigning military women. AR 600-13 states,

The Army's assignment policy for female soldiers allows women to serve in any officer or enlisted specialty or position except in those specialties, positions, or units (battalion size or smaller) which are assigned a routine mission to engage in direct combat, or which collocate routinely with units assigned a direct combat mission.[4]

---

[3] Headquarters, U.S. Department of the Army (1992).

[4] Headquarters, U.S. Department of the Army (1992, p. 1).

Important to understanding the Army policy is recognizing that it defines direct combat differently from the DoD policy. The Army policy defines direct combat as follows:

> Engaging an enemy with individual or crew served weapons while being exposed to direct enemy fire, a high probability of direct physical contact with the enemy's personnel and a substantial risk of capture. Direct combat takes place while closing with the enemy by fire, maneuver, and shock effect in order to destroy or capture the enemy, *or while repelling the enemy's assault by fire, close combat, or counterattack.*[5]

This definition of direct combat is different from the definition provided in the Aspin memorandum. The Army definition adds the requirement for a substantial risk of capture. Additionally, and very importantly, the Army policy includes "repelling the enemy's assault by fire, close combat, or counterattack" in its definition of direct combat.

There are several important differences between the Army and the DoD policies. First, the DoD policy restricts the assignment of women to units whose *primary* mission is direct ground combat, whereas the Army restricts assignment to units that have a *routine* mission of direct combat. Second, the Army also restricts assignment to units that collocate with direct combat units. Third, the Army and DoD policies define *combat* differently: The Army's definition of *direct combat* includes a requirement that there be a risk of capture, but also includes "repelling the enemy's assault." These differences are significant, and it is notable that the Army did not update its policy when Congress repealed the legal restrictions against women serving in combat aircraft position and on combatant ships nor when Aspin revised the DoD policy in 1994. The implications of these differences are discussed later in this monograph in the context of operations in Iraq.

---

[5] Headquarters, U.S. Department of the Army (1992, p. 5). Emphasis added.

Applying the Assignment Policies

It is important to understand that these policies are assignment policies pertaining to women rather than general employment policies. The policies provide guidance about the specialties, positions, and units to which women can be formally assigned. However, the policies do not constrain what individual women can do in operations. On the contrary, the Army policy explicitly states that, once properly assigned, female soldiers are subject to the same utilization policies as their male counterparts. The Army uses this policy as the basis for assigning women and implements those assignments in both the active Army and the reserve component with the Direct Combat Probability Code (DCPC) system, which uses the following three dimensions to classify each Army position: (1) the duties of the position and the area of concentration or military occupational specialty (MOS), (2) the unit's mission, and (3) routine collocation.

The extent to which a unit's activities in Iraq are relevant to this assessment of the assignment policy differs for the DoD and the Army policies. The DoD direct combat restriction focuses on the primary mission of direct combat units. Thus, the doctrine pertaining to the units, not their activities in theater, will determine the units to which women can be assigned. The Army policy, however, includes restrictions that require an assessment of the units' activities. These restrictions pertain to collocation as well as involvement in direct combat. In the case of the direct combat restriction, the Army policy precludes women from being assigned to a unit whose routine mission includes direct combat. Because the routine activities of a unit might change without a corresponding change in doctrine, it is important to assess unit activities in the theater. The focus on assigning women rather than individually utilizing women is discussed in more detail in Appendix B.

Opportunities Available to Army Women

As a result of changes in the DoD assignment policy for military women, additional Army units and occupations were opened to women.[6] In their January 12, 1994, memorandum to the Secretary of Defense, the Secretary of the Army and the Army Chief of Staff stated the intent to open the following units and occupations to women:

- maneuver brigade headquarters
- division military police companies
- chemical reconnaissance and smoke platoons
- mechanized smoke companies and smoke platoons
- divisional forward support battalions (FSBs) (forward maintenance support teams)
- engineer companies (medium girder bridge and assault float bridge)
- military intelligence collection and jamming companies
- Washington, D.C.–area ceremonial units.[7]

The Army's implementation of the changed DoD policy also resulted in opening positions within the headquarters of some maneuver and separate brigades, as well as in other types of units, such as the special forces group.[8]

As of the end of fiscal year 2006, the active-component Army includes over 48,000 women, who have the opportunity to serve in

---

[6] The Army assignment policy, AR 600-13 (Headquarters, U.S. Department of the Army, 1992), predates the Aspin memorandum. Positions were opened to Army women in 1994 as a result of Secretary Aspin's removal of the risk rule from the DoD assignment policy.

[7] Gordon R. Sullivan, U.S. Army Chief of Staff, and Togo D. West, Jr., Secretary of the Army, "Direct Combat Definition and Assignment Rule," memorandum to the Secretary of Defense, January 12, 1994.

[8] The Army assignment policy, AR 600-13 (Headquarters, U.S. Department of the Army, 1992), did not change with the change in DoD policy.

92.3 percent of Army occupations; 70.6 percent of Army positions are open to women.[9]

Basis for the Current Study

The Army has recently changed its organizational structure to a modular one that involves a different command structure and form of interaction between maneuver and support units. Army units, including brigade combat teams (BCTs), form the new modularized structure characteristic of Iraq deployment. Women have been an integral part of this structure, comprising approximately 10 to 20 percent of Army personnel deployed to Iraq and participating in almost every kind of unit or subunit open to women within BCTs.[10]

Iraq also presents a different kind of warfare. The assignment policy was drafted at a time when battles were assumed to be linear, characterized both by a front line, where direct contact with the enemy occurred, and relatively safer areas in the rear. In Iraq, U.S. forces confront an asymmetric threat. In other words, rather than fighting an enemy that uses similar weapons and techniques, U.S. forces confront an enemy that attempts to harm U.S. assets without going up against the "teeth" of U.S. defenses. For example, counterinsurgents in Iraq have been more likely to target unarmored convoys or civilian locations than better-armed and -defended systems, such as the Abrams tank or the Bradley fighting vehicle. Additionally, the asymmetric warfare in Iraq is occurring on a nonlinear battlefield.[11]

Given the Army's recent modularization, as well as the differences between military missions in Iraq in the context of the global war on terrorism (GWOT) and military missions fought on the linear battle-

---

[9]   Data provided by Office of the Deputy Chief of Staff for Personnel (Army G-1), Women in the Army office. See Appendix C for additional detail on Army occupations open and closed to women.

[10]   Our analysis of the roles in which Army women deployed to Iraq is included in Appendix D.

[11]   Asymmetric threats and nonlinear battlefields are discussed in more detail in Appendix G.

fields of past military engagements, concerns have arisen among some members of Congress and other interested parties as to whether the Army's use of women in Iraq is consistent with existing policies.

In May 2005, House Armed Services Committee Chairman Duncan Hunter and Military Personnel Subcommittee Chairman John McHugh cosponsored an amendment to the National Defense Authorization Act for Fiscal Year 2006 that would have made into law the 1994 assignment policy and would have also precluded DoD from opening new positions to women without an act of Congress.[12] Democrats and active and retired military leadership resisted the measure, stating that it would "tie the hands of military commanders in a time of war" and undermine the recruiting, morale, and careers of professional military women.[13] Subsequently, Chairman Hunter proposed a revised amendment to require the Secretary of Defense to give 60 days' (instead of the prior 30 days') notice to Congress before changing the assignment policies for women or opening or closing new positions to women and to report whether DoD was currently complying with the 1994 policy.[14]

While the final law did not change the reporting requirement, Section 541 of Public Law 109-163, January 6, 2006, does require an investigative report:

> Not later than March 31, 2006, the Secretary of Defense shall submit to the Committee on Armed Services of the Senate and the Committee on Armed Services of the House of Representatives a report of the Secretary's review of the current and future implementation of the policy regarding the assignment of women

---

[12] See, for example, Liz Sidoti, "House Committee Votes to Ban Women in Combat," *Capitol Hill Blue*, May 19, 2005a.

[13] See, for example, Ann Scott Tyson, "More Objections to Women-in-Combat Ban," *Washington Post*, May 18, 2005, p. A5. See also U.S. Senate, S 1134-IS, To Express the Sense of Congress on Women in Combat, 109th Congress, 1st Session, May 26, 2005, which was introduced by Senators Hilary Rodham Clinton, Susan Collins, Mary L. Landrieu, Patty Murray, Jack Reed, and Barbara A. Mikulski.

[14] U.S. House of Representatives, House Armed Services Committee, "Hunter Statement on Department of Defense Direct Ground Combat Policy," press release, Washington, D.C., May 25, 2005c.

as articulated in the Secretary of Defense memorandum, dated January 13, 1994, and entitled, "Direct Ground Combat Definition and Assignment Rule." In conducting that review, the Secretary shall closely examine Army unit modularization efforts, and associated personnel assignment policies, to ensure their compliance with the Department of Defense policy articulated in the January 1994 memorandum.

Subsequently, the Under Secretary of Defense for Personnel and Readiness communicated to Congress the need to extend the deadline past March 2006 and informed Congress that the RAND Corporation had been engaged to assist in data collection and analysis.[15] Accordingly, this monograph is intended to provide the Office of the Secretary of Defense (OSD) with analysis it may consider in its response to Congress.

Objectives and Scope of This Study

This OSD-sponsored study was designed to assess whether there is a common understanding—a shared interpretation—of the assignment policy; to determine whether, given Army operations in Iraq, the Army is currently complying with policy; and to assess whether the policy is appropriate to the new military environment, evidenced by current operations in Iraq. This study focused particularly on the Army BCTs that deployed to Iraq in a modular configuration, with specific attention to the new organic relationships with brigade support battalions (BSBs). The intent of this research is not to prescribe policy, but rather to report research findings about the assignment policy, given Army operations in Iraq, and to identify issues for DoD's consideration in decisionmaking concerning policy, doctrine, and employment.[16]

---

[15] Personal communication from David S. C. Chu, Under Secretary of Defense, to Senator John W. Warner, chairman, Senate Armed Services Committee, and Representative Duncan Hunter, chairman, House Armed Services Committee.

[16] The scope of our study did not include an assessment of the appropriateness of the assignment policy to past operations nor did it consider operations in Afghanistan. Such investiga-

Although this monograph directly references DoD policy as articulated in the Aspin memorandum, Public Law 109-163 also requires the Secretary of Defense to inform Congress of any changes to the ground combat exclusion policy, in which

> the term "ground combat exclusion policy" means the military personnel policies of the Department of Defense *and* the military departments, as in effect on October 1, 1994, by which female members of the armed forces are restricted to units and positions below brigade level whose primary mission is to engage in direct combat on the ground.[17]

This passage is important because it suggests that the Army cannot choose not to adhere to its own policy and because the law specifically references the personnel policies of both DoD and the military departments. Thus, this monograph addresses both the DoD and Army assignment policies and identifies the manner in which they differ.

This effort focused on Army BCTs and their support units operating in the Iraqi theater. While our observations may apply to the other military services and to other ongoing or future operations, this study did not specifically address how the other services assign women, nor did it encompass operations by the Army or other services in Afghanistan.

Approach and Methodology

To accomplish the study objectives, this effort included three primary research tasks. The first task involved describing the assignment policy and establishing the perceived objectives of the assignment policy. The second task analyzed the Army's transformational modular combat and combat support design, function, and doctrine to determine whether the doctrine is consistent with the assignment policy. This study was developed with the recognition that the design, function, and doctrine

---

tions would be interesting for further research.

[17] Public Law 109-163, Section 541 (2006). Emphasis added.

of modular Army units were likely being adapted to the Iraqi theater. Accordingly, we placed more emphasis on the importance of the third research task: understanding how Army BCTs and BSB support units were employed in Iraq, the roles that were filled by women, and to what extent the assignment policy was both appropriate, given Army operations in the Iraqi theater, and complied within that context.

These tasks employed different research methods. We reviewed the relevant literature and debate and conducted 11 qualitative expert interviews with Army, OSD, and Joint Staff (JS) leadership to assess the objectives of an ideal assignment policy, to assess the extent to which the current policy meets those objectives, and to ascertain the extent to which they agree upon the meaning of the policy. This effort also included five interviews with congressional members and staff to discuss the objectives or intent of the policy.

This research also included qualitative interviews and focus groups with service members returned from Iraq. These interviews and focus groups were conducted at a schoolhouse and a unit installation. The Army selected the unit, based on the scheduled return and availability of units, and identified local officers to recruit the focus group and interview participants. In general, battalion and brigade command personnel were interviewed, and more junior personnel participated in focus groups, though a similar protocol was used in both interviews and focus groups. In total, 80 people from the two locations participated in 16 focus groups and eight individual interviews. The confidential 60-minute sessions were led by experienced RAND researchers.

The interview and focus group data were transcribed, coded, and analyzed using the grounded theory method with qualitative analysis software, which permitted the research team to identify themes in experiences of Army personnel in Iraq.[18] These interactions with recently returned personnel were extremely important to this effort because they informed the research team's understanding of "how things really worked" in Iraq. Many quotes and observations from these partici-

---

[18] For more information about the types of personnel who were interviewed and participated in the focus groups and the semistructured protocols used for these sessions, see Appendix F.

pants are included in this monograph; these participants are referred to as recent returnees to protect their identities while distinguishing their comments from those of senior Army, OSD, and JS interviewees.

Importantly, while we obtained a range of views from individuals in a variety of occupations and units, our interview and focus group data are not representative. Thus, they cannot be considered indicative of the extent to which observed practices are occurring in Iraq, nor can the absence of an observation be construed as evidence that such practices do not happen in Iraq. These observations are intended, instead, to indicate practices that occur at least among some units in Iraq and to indicate perspectives that are held by at least some returned service members. Further, in some cases, we chose to portray the range of perceptions provided by our participants, which is another reason the perceptions and attitudes reported in this monograph cannot be assumed to be predominant views. Additionally, although we indicate whether the comments were made by returned service members or during the senior interviews, we do not provide the personal characteristics of the returned service member who made each comment to protect their confidentiality.

This research also included a review of lessons learned and other materials provided by the Center for Army Lessons Learned, as well as discussions with other Army experts.

Organization of This Monograph

This monograph is comprised of five chapters, including this one. Chapter Two discusses whether the current assignment policy is understandable and describes the central objectives of an assignment policy. Chapter Three considers whether the Army is complying with its policy, given its operations in Iraq, without questioning the policy itself. Chapter Four discusses whether the language and concepts in the current assignment policy are appropriate for the new military environment and the Army's new structure. Operations in Iraq are considered reflective of the new military environment. The final chapter offers conclusions and recommendations.

This monograph also includes supporting appendixes. Appendix A includes the Aspin memorandum that is DoD policy. Appendix B discusses the difference between an assignment policy and an employment policy. Appendix C includes additional information about the opportunities available to current Army women. Appendix D describes the quantitative analysis of Army women deployed in Iraq. Appendix E includes more information about the interviews with Army, OSD, and JS leadership and congressional members and staff and provides more explanation of the objectives discussed during those interviews. Appendix F presents additional information about the interviews and focus groups conducted with Army personnel recently returned from Iraq. Appendix G discusses the Army's modularity and today's asymmetric warfare on the nonlinear battlefield in Iraq. Appendix H summarizes and describes the characteristics of Army women who have received the Combat Action Badge (CAB).

CHAPTER TWO

Is There a Shared Interpretation of the Assignment Policy for Army Women?

A policy crafted for a linear battlefield and threat uses words that are not appropriate for the characteristics of military operations today. As a result, problems arise in translating and determining the objectives of the policies. Therefore, this chapter discusses the perceived meaning and objectives of the assignment policy as they apply to women in the Army. We assess whether the precise prescription, or the "letter" of the current assignment policy, is understandable and then discuss whether the purpose, intent, or "spirit" of the policy is discernable. This analysis is based on interviews with Army, OSD, and JS personnel, as well as with members of Congress and their staff. It also includes interviews with Army personnel who have recently returned from Iraq. Through these interviews and other sessions with Army, OSD, and JS personnel, it became apparent that many of the terms in the assignment policy are now much less meaningful.[1]

Understanding the "Letter" of the Policy

The Wording of the Policy

There are several words and phrases in the assignment policy that are less meaningful in the context of the GWOT and, as addressed in

---

[1] The interviews with senior Army, OSD, and JS personnel are described in greater detail in Appendix E. The interviews and focus groups with recently returned service members are described in more detail in Appendix F.

this study, operations in Iraq. Such terms include *enemy* and related phrases. As noted in Chapter One, the DoD policy states that "direct ground combat is engaging an enemy on the ground" and that it "takes place well forward on the battlefield while locating and closing with the enemy."[2] The Army policy definition of direct combat includes "engaging an enemy," "being exposed to direct enemy fire," and "closing with the enemy."[3] Terms such as *enemy* and positional concepts such as *well forward* (used in the DoD policy) and *collocate and remain* (used in the Army policy) are now less meaningful. When interviewing senior Army, OSD, and JS personnel, we inquired about the meanings of some of these words and phrases.

The Enemy. When asked whether *enemy* had any meaning in the context of operations in Iraq, some interviewees felt that they were able to identify the enemy, but their definitions were based on the actions or intent of others. For example, the following was typical of these responses: "I don't think it's difficult; anyone that's shooting at me is an enemy." But interviewees generally agreed that, if someone was not actively shooting, it was difficult to ascertain the identity of the enemy. Other comments included the following:

> Unless they're shooting at you, you don't know.

> It is very difficult because we're not engaged with a regular, uniformed force. The tactics of IEDs [improvised explosive devices] and mortars makes it even more difficult to define. Who is planting the IEDs?

> The enemies no longer wear uniforms. They may be your friend one day and firing bullets the next. . . . There's no definition of *enemy*, and it's a changing concept.

One participant acknowledged the difficulty of defining the term *enemy* in the assignment policy but maintained that it was still important to include the concept of the enemy in the assignment policy:

---

[2] Aspin (1994).

[3] Headquarters, U.S. Department of the Army (1992, p. 5).

They are the enemies if they are working against the objectives of the nation and the Army. The ability to clearly delineate all forms of enemy is extremely difficult. There are different forms of enemy: large enemy formations, smaller formations but still in uniform, radical groups (that are not military), and possibly host-nation civilians that are anti-American. It's not all black and white, and *enemy* is not a clear term. But do not remove the term from the definition; we just need a cognitive appreciation that who the enemy is has changed.

There was greater consensus among personnel recently returned from Iraq, who tended to agree that anyone could be the enemy and that, often, everyone is considered a potential enemy. Consequently, they believe that everywhere in theater is dangerous, as conveyed by the following:

The enemy has no face.

Everyone could be a potential enemy.

Anyone. A woman, a child, a man, and old man. Anyone.

It could be your translator.

We also asked the senior interviewees whether, given the uncertainty of identifying the enemy, it was possible to know when U.S. forces were "closing with the enemy." This concept is important to the assignment policy because both the DoD and Army definitions of direct combat are partly based on the concept of closing with the enemy. All six senior Army participants and one of the JS/OSD participants provided definitions for *closing with the enemy*. While some participants viewed this concept as more clear cut than did others, several felt that it was relatively evident when a force was closing with the enemy, and one interviewee made the assertion that U.S. forces were at their least vulnerable when they were closing with the enemy:

Maneuver against the enemy to be in a position to destroy them if necessary.

Going out with specific information to capture or kill—that's "closing."

You are at your least vulnerable when you are closing with them. They're overmatched and it's almost an unfair fight.

Given the uncertain identity of the enemy, there was some inherent uncertainty about whether forces are in danger from hostile fire. This concept is included in the DoD policy as "being exposed to hostile fire" and in the Army policy as "being exposed to direct enemy fire." The Army leadership interviewed tended to assert that "a commander's read of the battlefield," "the intelligence preparation," or "intellectual preparation" of the battlefield informed the commander about whether personnel were being exposed to hostile or enemy fire. Nonetheless, there was uncertainty expressed in the interviews, with some participants noting the potential for hostile fire any time soldiers leave the forward operating base (FOB) and others acknowledging the potential even inside the FOB:

> We [the support element] were 15 to 25 kilometers behind the front line in the 1992 linear battlefield, doctrinally. It [the assignment policy] made sense then. Today, a soldier on the FOB who returns fire is in direct combat.

Forward and Well Forward. The concepts of *forward* and *well forward* were generally acknowledged to be almost meaningless in the Iraqi theater. Indeed, only half of the Army leadership interviewees and fewer than half of the senior OSD and JS interviewees provided definitions, while the others argued that these terms had no meaning. The definitions that were provided ranged from "in Iraq" to "where ground combat will be."

Returned service members also frequently explained that the assignment policy was based on a linear battlefield, which does not exist in Iraq. They tended to make comments such as

> There's no clear-cut front line. Some people do more dangerous stuff, but in my opinion, it's all the same. You could stay in the

road and be blown up. It's the way FSCs [forward support com-
panies] are.

There's no real clear definition [of combat exclusion] because it's
not a linear battlefield. No one can define what they're excluding
us from. There's no line, especially when with the BCTs.

As one officer simply stated, "As soon as the plane takes off from
Kuwait, you're forward." Many participants referred to everywhere in
Iraq as forward or to *forward* being indefinable in Iraq. There were
some returned service members who believed that the concept of *forward* still existed, even if it was difficult to define. They typically provided conceptual answers, such as "Forward is directly proportional
to enemy activity." Another discussion group related the definition to
available amenities:

Participant 1: *Far forward* means a small FOB. The small patrol
bases were far forward. They didn't have water and they were
eating MREs [meals, ready to eat].
Participants 2 and 5: Yeah, it would be based on the conditions
at the base.
Participant 3: Because conditions and risk are intertwined.
Participant 2: That's because there's not as many people there to
repel any risk.

However, most participants were dubious that *far forward* had
any meaning in Iraq. As one recent returnee expressed, "There's no
battle line. There's no 'they're here and we're here.' It's not like that."
The linear concepts of *forward* and *rear* do not have meaning in Iraq;
instead, personnel in Iraq are subject to attacks by relatively small units
oriented from many directions. This is reflected in and consistent with
DoD acknowledgement of the broad threat and lack of safe areas in
Iraq, as well as in the Army tactics, techniques, and procedures (TTPs),
discussed earlier, requiring all units to prepare to defend against these
kinds of threats.

Collocate and Remain. An understanding of the concept *collocate*
is central to understanding the assignment policy for military women.

The Army policy states that units are closed to women if they "physically collocate and remain" with those units whose doctrinal mission is direct combat. However, there is some uncertainty about the meaning of *collocate*. Most senior interviewees and recent returnees pronounced the word /koh`loh'kayt/ and appeared to refer to the definition of *collocate* as "to place two or more units in close proximity so as to share common facilities."[4] We have hereafter referred to that usage as *colocate* or *colocation*. In contrast, a small number of interviewees distinguished *collocate* (/ka lE ket/) from *colocate*, with a different pronunciation as well as a different definition. These individuals explained that *collocate* implies a high level of interaction and interdependency between the units, rather than just physical proximity. By this definition, *collocate* means to "place in the proper order"[5] or "to occur in conjunction with something,"[6] suggesting that neither could perform its mission without the other. One returned service member explained the distinction as follows:

> If we're going to pass an FSC area where they are supporting maneuver units, we would often deliver it to them, especially the fuel. So I'll say that when we did that, we were colocated, but not collocated, because we weren't intermingled.

While both the Army and DoD policies use the word *collocate*, the Army policy still appears to define the word as *colocate*, with a definition that mentions only proximity and not interaction or proximate interdependence:

> Collocation [o]ccurs when the position or unit routinely physically locates and remains with a military unit assigned a doctrinal mission to routinely engage in direct combat. Specifically, positions in units or sub-units which routinely collocate with units assigned a direct combat mission are closed to women. An entire unit will not be closed because a sub-unit routinely collocates

---

[4]   *Webster's New Collegiate Dictionary*, Springfield, Mass.: G & C Merriam Co., 1979.

[5]   *Wordsmyth English Dictionary-Thesaurus*, Wordsmyth, updated continuously.

[6]   *Webster's New Collegiate Dictionary* (1979).

with a unit assigned a direct combat mission. The sub-unit will be closed to women.[7]

The difference between *collocate* and *colocate* is subtle but important. In this monograph, we will consider both definitions related to operations in Iraq in Chapter Three, when we assess whether Army operations in Iraq comply with the assignment policy.[8]

Related to the definition of *collocate* is the meaning of *remain*. We also asked the senior interviewees for their definition of *remain* as it is used in the assignment policy. Six of nine answers indicated that the definition was operation- or situation-dependent and thus could not be defined comprehensively. Two interviewees were unable to define the term. Of the time-based answers we received, they ranged from "one minute" to "a minimum of 24 hours." Of note, one senior interviewee expressed concern about the potential for abuse or manipulation of any time-based policy.

Discerning the "Spirit" of the Policy

Given some confusion about the wording, or "letter," of the policy, it is important to ascertain whether the "spirit" of the policy is discernible. In other words, is there broad understanding of what the policy is trying to achieve, despite some uncertainty about the specific vocabulary or the application of the policy to a new kind of theater and a new Army organizational structure?

Our research suggests that the policy objectives are not clear. For example, the policy articulates that women are not to be assigned to ground combat units, but it does not specify why this is important. Comments from some popular media outlets and from some members of Congress suggest that the purpose of the policy is, or should be, to protect women. This was apparent in remarks by Harold Stavenas,

---

[7]   Headquarters, U.S. Department of the Army (1992, p. 5).

[8]   In this monograph, we generally use *collocate* when referring to interdependency and *colocate* when referring to proximity.

communications director of the House Armed Services Committee, who said, "We are exposing women to combat more than we should."[9] Advocacy groups and pundits have inquired, "Do we really want America's mothers on the frontlines of the War on Terror?"[10] "Why were the [American] women so vulnerable to capture and likely abuse as prisoners?"[11] and "Engaging the enemy in this uncivilized thing we call war is a job for men, not women."[12] Additionally, they have asserted that the previous risk rule "reflected the prevailing view that female soldiers should not be needlessly exposed to risk of capture while serving in close proximity to close combat units such as the infantry, armor, and field artillery."[13] The popular media have also reported that "there is a deeply rooted belief that women should be protected rather than protectors," but Heather Wilson, a New Mexico congresswoman and former military officer, asserts that "Americans have accepted that women make all kinds of contributions . . . including protecting this country from its enemies."[14] Likewise, another expert on military women asserts that "the public accepts that women are in the military, that there are going to be shootings, and that they're going to be dying, and that's fine—with most people."[15] Disagreement about whether the policy is designed to protect women was also conveyed in private sessions between the RAND research team and members of Congress.

---

[9] Luis R. Agostini, "Women's Combat Support Role Could End in Iraq," *Marine Corps News*, May 19, 2005.

[10] "Do We Really Need Mothers in Combat? Support Amendment to Uphold Ban on Women in Combat," *Eagle Forum*, May 13, 2005.

[11] Center for Military Readiness, "Women in Land Combat," Livonia, Mich., report no. 16, April 2003a, p. 1.

[12] Kate O'Bierne, Washington editor of the *National Review*, quoted in Sharon Cohen, "Women Take on Major Battlefield Roles," Associated Press, December 3, 2006.

[13] Center for Military Readiness (2003a, p. 2).

[14] Dave Moniz, "Female Amputees Make Clear That All Troops Are on the Front Lines," *USA Today*, April 28, 2005.

[15] Lory Manning, retired Navy captain and director of the Women in the Military project at the Women's Research and Education Institute, in Cohen (2006).

Academics and others, such as Anna Simons, suggest that the safety of male service members is at issue by arguing that the presence of women among ground combat personnel breeds distraction, dissension, and distrust, undermining unit effectiveness and, thus, the safety of men. Indeed, the same author argues that the physical weakness of women would preclude women from carrying to safety their wounded male colleagues who were put at risk by women's very presence.[16]

Still others would agree with Representative Susan Davis' assertion: "The underlying spirit of the current policy is that women are to be afforded increasing opportunities within the military."[17]

Given these myriad views, the interviews with senior Army, OSD, and JS leadership also gathered perspectives about their perception of an ideal assignment policy and the extent to which the current policy satisfied those objectives. Similar but less structured interviews were conducted with selected members of Congress.[18] Each Army, OSD, and JS senior interviewee was asked to assess the importance of each in a list of possible objectives that have emerged from the ongoing discussion of women in the military. The policy objectives included the following:

- Maximize the operational effectiveness of the military.
- Maximize flexibility in assigning people.
- Maintain current opportunities for women.
- Open new career opportunities for women.
- Provide career opportunities to make women competitive with their male counterparts in career advancement.
- Protect female service members from physical harm.
- Protect male service members from physical harm by excluding women from ground combat.

---

[16] Anna Simons, "Women in Combat Units: It's Still a Bad Idea," *Parameters, U.S. Army War College Quarterly*, Summer 2001, pp. 89–100.

[17] U.S. House of Representatives, House Armed Services Committee, "McHugh/Hunter Provision Limits Flexibility of Commanders During War," press release, May 24, 2005b.

[18] Appendix E provides more information about the interviews with senior personnel, including the protocol and the precise wording and an explanation of the objectives.

- Simplify unit leadership by limiting male-female interaction.
- Exclude women from ground combat occupations and units.
- Exclude women from occupations that require considerable physical strength.
- Ensure buy-in from all involved parties/stakeholders through compromise.

In each case, interviewees were asked to indicate their level of agreement with a statement, for example, "It is important that an assignment policy for women maximize the effectiveness of the military." Possible answers ranged from "strongly agree" to "strongly disagree." Once through the entire list, the participants were then asked to assess whether each objective was reflected in the current assignment policy.[19]

Table 2.1 presents the responses of the five OSD and JS participants for each objective. The objectives are listed in the first column, and the left half of the table body indicates the number of interviewees who felt that an objective was important in an assignment policy for military women. For example, all interviewees strongly agreed that it was important that an assignment policy maximize the operational effectiveness of the military, and all five of the participants either strongly agreed or agreed that an assignment policy should maximize flexibility in assigning people. Only three of the OSD and JS participants felt that an assignment policy should maintain current opportunities for military women, while the other two interviewees reflected discontent or neither agreed with nor disagreed with the importance of maintaining current opportunities. Consistent with that, three believed that an assignment policy should open new opportunities for military women. Of note is that responses were evenly split about whether the assignment policy should protect women from physical harm (of four responses), and nonneutral responses were also evenly

---

[19] The objectives "Maintain current opportunities for women" and "Open new career opportunities for women" were excluded from the second set of questions addressing whether the current assignment policy successfully addresses those objectives, since the questions do not make sense in the context.

Table 2.1
OSD and JS Senior Interviewee Perspectives on Assignment Policy Objectives

Objective	Important in Assignment Policy					Reflected in Current Assignment Policy				
	SA	A	N	D	SD	SA	A	N	D	SD
Maximize operational effectiveness of military	5						4			
Maximize flexibility in assigning people	4	1				1	3			
Maintain current career opportunities for women	1	2	1	1		n/a	n/a	n/a	n/a	n/a
Open new career opportunities for women		3	2			n/a	n/a	n/a	n/a	n/a
Provide career opportunities to make women competitive with men	1	4				2		2		
Protect female service members from physical harm		2		1	1				3	1
Protect male service members from physical harm		1	2		1	1	1	2		1
Simplify unit leadership by limiting male-female interaction			2	2					3	2
Exclude women from ground combat occupations/ units	2	2				3	1	1		
Exclude women from occupations requiring considerable physical strength		1		3					5	
Act of compromise		2	1	1					4	1

NOTE: SA = strongly agree. A = agree. N = neither. D = disagree. SD = strongly disagree. n/a = not applicable. For each of the two questions, the maximum number of responses is five; however, not all respondents answered each question.

split about whether an assignment policy should protect male service members (of four responses).

The right half of the table body indicates the extent to which these same OSD and JS senior interviewees felt that these objectives were reflected in the current assignment policy. Several observations emerge from a comparison of the two sides of the table body. First, while all the interviewees felt that it was important for an assignment policy to provide career opportunities to make women competitive with men, they were split on whether the current policy did so (of four responses). None of the interviewees felt that the current policy protected women or simplified unit leadership by limiting male-female interaction, and only one believed that the current policy protected men. Likewise, none of them believed that the current policy successfully kept women from occupations requiring considerable physical strength.

Table 2.2 builds on Table 2.1 and compares the six Army perspectives to the five OSD/JS perspectives. Table 2.2 displays both the responses from the OSD and JS senior interviews and the responses from the senior Army interviews about the objectives that should be maximized in an assignment policy and how much the current assignment policy maximizes those objectives. For each objective, the Army responses appear above the OSD/JS responses. Overall, the Army answers were more similar to the OSD/JS responses than they were different regarding the objectives that were or were not important in a prospective assignment policy.

In sum, both the Army and the OSD/JS interviewees agreed that it was important for a policy to maximize operational effectiveness, maximize flexibility in assigning people, maintain current opportunities for women, provide career opportunities to make women competitive with male service members, and exclude women from ground combat. They agreed that it was not important for a policy to protect female service members, to protect male service members, to simplify leadership, or to be an act of compromise. OSD/JS interviewees were slightly more inclined than were Army interviewees to emphasize providing new opportunities for women. There was considerably less consensus among Army and OSD/JS interviewees regarding the objectives

Table 2.2
Army, OSD, and JS Senior Interviewee Perspectives on Assignment Policy Objectives

Objective	Respondent	Important in Assignment Policy					Reflected in Current Assignment Policy				
		SA	A	N	D	SD	SA	A	N	D	SD
Maximize operational effectiveness of military	Army	4	2				1	2	2	1	
	OSD/JS	5						4			
Maximize flexibility in assigning people	Army		6					1	2	3	
	OSD/JS	4	1				1	3			
Maintain current career opportunities for women	Army	1	4	1			n/a	n/a	n/a	n/a	n/a
	OSD/JS	1	2	1	1		n/a	n/a	n/a	n/a	n/a
Open new career opportunities for women	Army		2	4			n/a	n/a	n/a	n/a	n/a
	OSD/JS		3	2			n/a	n/a	n/a	n/a	n/a
Provide career opportunities to make women competitive with men	Army	2	4				2	1	2	1	
	OSD/JS	1	4					2		2	
Protect female service members from physical harm	Army			1	4	1			2	4	1
	OSD/JS		2		1	1				3	1
Protect male service members from physical harm	Army		2	1	3				1	4	1
	OSD/JS		1	2		1		1	1	2	1

Table 2.2—Continued

Objective	Respondent	Important in Assignment Policy					Important in Current Assignment Policy				
		SA	A	N	D	SD	SA	A	N	D	SD
Simplify unit leadership by limiting male-female interaction	Army		1		4					7	
	OSD/JS			2	2					3	2
Exclude women from ground combat occupations/ units	Army		3	1	1		4		1	1	
	OSD/JS	2	2					3	1	1	
Exclude women from occupations requiring considerable physical strength	Army		3	1	2			1	3	2	
	OSD/JS		1		3					5	
Act of compromise	Army		1	2	3				1	3	2
	OSD/JS	2	2	1	1					4	1

NOTE: SA = strongly agree. A = agree. N = neither. D = disagree. SD = strongly disagree. n/a = not applicable. For each of the two questions, for OSD and JS staff, the maximum number of responses is five; however, not all respondents answered each question. For each of the two questions, for Army leadership, the maximum number of responses is seven; however, not all respondents answered each question.

that are reflected in the current policy. There was a majority agreement that the current policy does exclude women from ground combat. There was not majority agreement regarding the extent to which the other objectives are reflected in the current policy.

The research team conducted fewer interviews with congressional members and staff, but the interviews suggest the lack of a consistent perspective on either the purpose of the existing assignment policy or the ideal objectives for a revised assignment policy. Indeed, when the interviews with members of Congress are considered in concert with the public statements of congressional members, the same degree of agreement seen among the DoD interviewees is not present. While some members answered similarly to the DoD interviewees, especially about the importance of military effectiveness, there tended to be a difference in opinion about whether the emphasis should be on expanding women's opportunities and about whether the policy should protect women.

Summary

Our interviews with senior personnel from the Army, OSD, and JS, as well as our sessions with personnel recently returned from Iraq, confirm that there is no shared understanding of the meaning of many of the words used in the DoD and Army assignment policies, including *enemy*, *forward* or *well forward*, and *collocation*. This is a result of the many changes that have occurred since the implementation of the policies, including the Army's transformation,[20] as well as the nature of warfare in Iraq.

Given the lack of a precise understanding of the "letter" of the policy, this research sought to ascertain whether there was a consistent understanding of the "spirit" of the policy. Senior Army, OSD, and JS personnel fairly consistently portrayed the objectives of an ideal policy. However, we find from interviews with and public statements of mem-

---

[20] The Army's transformation was mentioned in Chapter One and is discussed in greater detail in Appendix G.

bers of Congress and interviews and meetings with congressional staff that there is not the same agreement in Congress about the intent of a prospective assignment policy and that there is not a common understanding or agreement about the objectives reflected in, or the "spirit" of, the current assignment policy.

Is the Army Complying with the Assignment Policy?

This chapter considers whether the Army is currently complying with the assignment policy for women. This chapter focuses on operations in Iraq, as portrayed in the sessions conducted with recently returned service members, to consider whether the Army is assigning women to units consistent with the assignment policy. There are three primary considerations in this assessment. First, are women assigned to direct combat units, i.e., maneuver units below the brigade level (or battalion size or smaller)?[1] Second, have the support units to which women are assigned gained a mission of direct combat? Third, are there women in units that are collocated with direct combat units? The first two of these considerations are relevant to both the DoD and Army assignment policies, whereas collocation is a constraint only in the Army policy.[2]

Are Women Assigned to Direct Combat Units Below the Brigade Level?

Both the DoD and Army assignment policies preclude women from being assigned to direct combat units below the brigade level. In

---

[1] The DoD policy specifies "below the brigade level," whereas the Army policy specifies "battalion size or smaller."

[2] As discussed earlier, although the DoD policy permits the Army to include a collocation constraint in its assignment policy, collocation is not a constraint in the DoD policy.

determining whether current assignment practices comply with that restriction, it is important to consider both whether women are being assigned to maneuver units, which have traditionally been recognized as direct combat units, and whether women have been assigned to support units that might have adopted a direct combat mission. Throughout this assessment, it is important to recall that the DoD policy refers to units with a primary mission of direct combat, whereas the Army policy precludes the assignment of women to units that have a routine mission of direct combat.

Are Women Assigned to Maneuver Units?

This analysis considers whether women are assigned to maneuver units whose established primary mission is direct combat.[3] Personnel are individually assigned to a unit. However, their entire unit can be assigned to another unit, altering—at least in its practical operational aspects—the prior chain of command. We found no evidence that individual women were being assigned to maneuver units, nor did we find any evidence that female personnel were in support units assigned to maneuver units below the brigade level. Instead, the command relationship between the FSC and the maneuver units appeared to be a direct support relationship.[4] Consistent with this, the most recent BCT

---

[3] The DoD policy restricts women from being assigned to units whose "primary" mission is direct combat. The Army restricts women from assignment to units with a routine mission of direct combat. Any unit with a *primary* mission of direct combat would routinely conduct that mission. Thus, this initial inquiry into whether women are assigned to units whose *primary* mission is direct combat addresses both policies.

[4] Support relationships define the purpose, scope, and effect desired when one capability supports another. Support relationships establish specific responsibilities between the supporting and supported unit. Army support relationships are direct support, general support, general support reinforcing, and reinforcing. *Direct support* generally implies a mission requiring a force to support another specific force and authorizing it to answer directly the supported force's request for assistance. This relationship is normally habitual, since it requires the synchronization of standard operational procedures and mutual understanding of each unit's capabilities and missions.

doctrine states that "FSC's are assigned to the BSB, but usually are OPCON to their supported battalions."[5]

However, some of the personnel interviewed felt that the FSC relationship with the maneuver unit was consistent with being assigned and that their identity and, in some cases, their loyalty, was primarily to the maneuver unit. There were also some instances in which the FSC personnel felt that their command relationship to the BSB was in word only, as conveyed in the following three comments:

> RAND: How did you interact with the maneuver battalion?
>
> Participant: We're part of the battalion. We do everything they task us to do.

> There are paperwork issues for the BSB, but the FSCs know that they belong to the maneuver battalion. The [FSC] soldiers don't care [that they are part of the BSB]. It's irrelevant to them. The

---

[5] Headquarters, U.S. Department of the Army (2006b). OPCON, or operational control, is defined in U.S. Department of Defense, *Department of Defense Dictionary of Military and Associated Terms*, Joint Publication 1-02, April 12, 2001 (as amended through January 5, 2007), p. 389, as follows:

> Command authority that may be exercised by commanders at any echelon at or below the level of combatant command. Operational control is inherent in combatant command. Operational control is inherent in combatant command (command authority) and may be delegated within the command. When forces are transferred between combatant commands, the command relationship the gaining commander will exercise (and the losing commander will relinquish) over these forces must be specified by the Secretary of Defense. Operational control is the authority to perform those functions of command over subordinate forces involving organizing and employing commands and forces, assigning tasks, designating objectives, and giving authoritative direction necessary to accomplish the mission. Operational control includes authoritative direction over all aspects of military operations and joint training necessary to accomplish missions assigned to the command. Operational control should be exercised through the commanders of subordinate organizations. Normally this authority is exercised through subordinate joint force commanders and Service and/or functional component commanders. Operational control normally provides full authority to organize commands and forces and to employ those forces as the commander in operational control considers necessary to accomplish assigned missions; it does not, in and of itself, include authoritative direction for logistics or matters of administration, discipline, internal organization, or unit training.

paperwork [goes through the formal channels to the BSB], but I wear the [maneuver] battalion T-shirt. I know where I belong.

Ask all the FSC soldiers who they belong to. Their guidon [unit marker flag] is blue. They'd all say [the maneuver unit number]. They look like us, smell like us, did everything we did.

Some service members felt that assigning women to the FSCs was a de facto violation of the assignment policy, as the following comments convey:

[The Army has] gotten around this legally because they are assigned. They are organic, assigned to the BSB. Yet they are attached [to the maneuver units] for every other purpose.

Because of females, we're assigned to the BSB on paper, but when deployed, all FSCs are with the infantry battalions. You don't see the BSB until UCMJ [Uniform Code of Military Justice]. If you have issues, you have to wait to get a convoy to go to the main FOB [where the BSB is located].

Thus, women are not assigned to maneuver units below brigade, and this complies with the DoD and Army assignment policies. However, we found that there are circumstances in which support units with women are in a relationship with maneuver units that is only very slightly different from being assigned, and that, in some circumstances, they have a closer relationship with the maneuver unit than with the unit in their assigned chain of command. Although these assignments meet the "letter" of the assignment policy, the assignments may involve activities or interactions that framers of the policy sought to rule out and that today's policymakers may or may not still want to rule out.

Have Support Units Adopted a Direct Combat Mission?

As discussed in Chapter One, the DoD and Army definitions of direct combat differ somewhat. They are repeated here for convenience. The 1994 Aspin memorandum provided the following definition:

Direct ground combat is engaging an enemy on the ground with individual or crew served weapons, while being exposed to hostile fire and to a high probability of direct physical contact with the hostile force's personnel. Direct ground combat takes place well forward on the battlefield while locating and closing with the enemy to defeat them by fire, maneuver, or shock effect.[6]

The Army's definition of direct combat omits mention of *ground* but does include a risk of capture clause and a reference to "repelling the enemy's assault":

Engaging an enemy with individual or crew served weapons while being exposed to direct enemy fire, a high probability of direct physical contact with the enemy's personnel and a substantial risk of capture. Direct combat takes place while closing with the enemy by fire, maneuver, and shock effect in order to destroy or capture the enemy, or while repelling the enemy's assault by fire, close combat, or counterattack.[7]

Because of the difficulty of defining *well forward* in current military operations, our analysis focuses on the activities of units rather than their relative locations on the battlefield. Thus, we considered whether Army women were in units that engaged the enemy as defined in the DoD and Army policies. First, were women assigned to units that participated in "locating and closing with the enemy"? We found that, while women were not assigned to such units, women did accompany male personnel who located and closed with enemy personnel. Specifically, women accompanied combat arms personnel on combat operations but did not participate in offensive operations directly:

On searches, we'd bring females [but] they'd stay in the Bradley until the building was cleared. No way would people find them in [an offensive combat situation]. Won't put them in an ambush

---

6   Aspin (1994).

7   Headquarters, U.S. Department of the Army (1992, p. 5).

[situation]. It's not their MOS. We wouldn't put mechanics in one either.

They were daily with teams kicking down doors. HUMINT [human intelligence] stands back until the building is secure. They do that whether male or female; they stand back. It's not part of their job to kick down the door.

Despite this close interaction, we found no evidence that women, even those trained to do so, were involved in initiating direct combat. Even one female soldier who had received training on how to clear a building acknowledged that she was not likely to do so:

I was trained to kick in doors. I know how to clear a building. . . . I was the only female out of 30. My job is to train other soldiers how to do it. . . . Once I walked in that class they were like, "Oh boy, we'll train you but you'll probably never use this," just based on my gender.

Further, when asked, commanders confirmed that they did not send women out to do combat patrols. Thus, women were not actively engaging and closing with the enemy, though they were interacting closely with teams that were doing so. While policymakers should be aware of this proximity, these activities appear to satisfy a strict reading of both the DoD and Army assignment policies. Thus, the support units to which women were assigned did not locate and close with the enemy or initiate direct combat.

Next, we consider whether women were assigned to units that "[engaged] an enemy with individual or crew served weapons while being exposed to direct enemy fire, a high probability of direct physical contact with the enemy's personnel and a substantial risk of capture," which is the wording from the Army's policy. Although we did not find, again, that women were assigned to units whose mission was to "[close] with the enemy by fire, maneuver, and shock effect in order to destroy or capture the enemy," we did find that support units engaged the enemy. Whether this engagement might be interpreted as "repelling the enemy's assault by fire, close combat, or counterattack" (as proscribed in Army policy) depends on the interpretation of this language and, more specifically, on the determination of whether self-defense activities are "repelling the enemy's assault."

Regarding this, we found consistent evidence throughout our interviews and focus groups that, while maneuver units sometimes provided facility or convoy security, support units were also trained and expected to provide their own convoy security. This emphasis on self-defense and agility is evident in the following comments from returned personnel:

> Participant 1: They've changed the TTPs for FSBs and convoys. The tactics changed. Before, the convoy was supposed to drive through [when fired upon], and now you are supposed to stop on the way [and engage the enemy].
>
> RAND: For FSBs?
>
> Participant 1: For convoys in general. If you're on the road [and come under fire], they want you to stop and engage the enemy and then proceed. . . . They are starting to train this in Kuwait because they decided that the prior practice of rolling through an attack was encouraging the enemy activity. They decided the convoy either needed to stop and discourage this or have a [combat arms] unit attached to do it.
>
> Participant 2: But even if the combat arms unit is protecting the convoy and fights back, the entire convoy stops and becomes involved. So this is true for convoys that are providing their own security and those that have external assets for security.

And from another discussion:

> The bottom line is that there are not enough troops. A field artillery battalion was covering an area that three battalions used to cover, so the FSC was doing route clearances.

Additional comments, including the following, confirmed that while support units were not initiating direct combat (or purposely closing with the enemy), self-defense was one of their routine missions practiced in the theater:

[The convoy] mission was not to go find anyone and shoot them. It was self-protection. That's part of our unit's mission. The METL [mission-essential task list] tasks include "repel a level I threat." Level I includes snipers and small teams. Anything above level I—that would need maneuver units to support. But repelling the enemy's attack by fire with crew-served weapons was definitely one of our missions in Iraq. There weren't enough maneuver units to provide all the convoy security we needed. . . . But engaging the enemy is not their primary mission. We do everything we can to deter first. The ROE [rules of engagement] and escalation-of-force rules in theater emphasize that you do everything you can to identify any potential threat and deter. You shine a light at them, show weapons, shine lasers, use flares. Any nonlethal means to show that you would rather not fire at them.

This emphasis on self-defense is consistent with the Army's 2006 Posture Statement, which details the Army's efforts to meet the demands of the new security environment by building a modular force based on BCTs and developing soldiers to serve as versatile warriors and adaptive leaders. For instance, all new recruits receive advanced training in marksmanship and live-fire convoy procedures, regardless of MOS, and current training involves tasks and drills that, in the past, were required only of infantry soldiers. Additional refinements have been instituted—and are still ongoing—to ensure that "Army leaders at all levels [will] be multi-skilled, innovative, agile, and versatile."[8]

Additionally, some support units appear to have even been conducting self-defense missions for other support units:

We were maintenance-focused but tasked in other ways. Our unit had one of the best success rates [for convoy security], so we were doing security for other units' convoys.

While some participants in support units reported that maneuver units had provided security for them, others had provided their own security, and some women participated in personal security teams

---

8 U.S. Department of the Army, "2006 Posture Statement," February 10, 2006, p. 15.

for the leadership. One commander explained that the circumstances varied across different units:

> RAND: Who had responsibility to secure the MSR [military supply route]?
>
> Participant: There was a battle space owner, but they didn't always sit along the road and guard. The convoy had to secure itself. The degree to which maneuver units provided security really varied with the different commanders. The maneuver commander might or might not emphasize securing the route over patrolling the inner city streets. Some would put tanks along the freeway and then others would not put anything there. Securing the route was not necessarily inherent in the maneuver missions.

Indeed, one commander described a convoy defense situation that earned a commendation for his young female soldier:

> There was a specialist who was about 20 years old. She was the driver of a 915. She was in a convoy and the truck in front of her gets destroyed. What we do is, when a convoy is hit, you get out of the hit area and form a box about two miles up the road. In this incident, there were IEDs, they blocked the road, there were snipers, RPGs [rocket-propelled grenades], and all at one time. So the convoy commander goes up the road to the box. The specialist is in the kill zone and the truck in front of her is on its side. We ran these convoys by having one Army truck and then three trucks driven by foreign nationals. So the sergeant, the convoy commander, tells her that he's going to come back. She says, "No, I'll bring them out." She earned the Bronze Star from that.

Another interpretation is that some form of enemy threat has always been anticipated for Army support units, including those that were not on the front lines, and thus all members of all units have always been expected to help the unit defend against level I threats—small bands of irregulars or terrorists conducting small raids, ambushes, or mortar or rocket attacks or setting booby traps—and that this has been true for decades. This logic—that support units and women would be involved in conflict—is consistent with the text of the Army response to the 1994 change in DoD policy, in which the Secretary of the Army states that

> the issue at hand is not one of deciding whether or not women will be "in combat." The nature of the modern battlefield is such that we can expect soldiers throughout the breadth and depth of a theater of war to be potentially in combat.[9]

This interpretation argues, likewise, that shared responsibility for a collective security mission has always been expected of personnel in support units and that the policy would not have prohibited women from being assigned to units that did so. Consistent with this, Army Field Manual 4-93.50, *Tactics, Techniques, and Procedures for the Forward Support Battalion*, states,

> Combat service support organizations are normally the units least capable of self-defense against a combat force. . . . However, *all units* must be able to defend against Level I activities (sniper, agents, saboteurs, or terrorist activities). They should be able to impede Level II attacks until assistance arrives.[10]

This kind of individual and group self-defense, or rear-area defense, against incidental small attacks, while perhaps not previously a *routine* mission, is not new to operations in Iraq. Further, the assignment policy is unlikely to have intended to preclude women from assignment to *all units*. Therefore, when the Army assignment policy includes "repelling the enemy's assault by fire, close combat, or counterattack," it may instead refer to the maneuver units' mission, which is conveyed with the same words as those of the assignment policy. For example, BCT doctrine states that the BCT's "core mission is to close with the

---

[9]   Togo D. West, Jr., Secretary of the Army, "Increasing Opportunities for Women in the Army," memorandum to the Under Secretary of Defense, Personnel and Readiness, July 27, 1994.

[10]   Headquarters, U.S. Department of the Army, *Tactics, Techniques, and Procedures for the Forward Support Battalion (Digitized)*, FM 4-93.50, Washington, D.C., May 2, 2002, p. 9-1. Emphasis added.

enemy by means of fire and maneuver to destroy or capture enemy forces, or *to repel their attacks by fire, close combat, or counterattack.*"[11]

However, several developments in Iraq might challenge this argument. One is the reports from some returned service members that support units provided security for other support units not because of a collective responsibility, but because of a particular support unit's proven proficiency in doing so. Thus, these units have gone beyond self-defense in providing security for other units. Additionally, a previously cited comment indicates that convoy security, in some cases, would stop and engage threats through which they previously might have passed, in an effort to discourage or prevent those attackers from harassing the next convoy. Whether these and other Army self-defense activities might be interpreted as going beyond traditional self-defense and whether they thus equate to "repelling the enemy's assault" are important considerations for policymakers and are key in determining whether the Army is in compliance with its assignment policy.

Beyond considering whether units adopted new missions that violate the proscription against direct combat, it is also important to consider that some units did not perform their established missions as might have been predicted based on their doctrine. Instead, returnees reported that they did not perform their METL tasks the way they had been trained because the doctrine was adjusted to accommodate the circumstances of the Iraqi theater. One FSC officer reported,

> They didn't have METL tasks to train on. There were no books yet for the FSC when we had just been stood up. No other units had an FSC. During the first year building the unit, they used [another unit's] documents. They used it as a task list to train on, but when you're out there, the battalion commander comes up and says, "I need this. Can you do it?" So whatever the battalion commander says, you make it happen.

Another returnee stated simply, "I'd say the doctrine actually failed us." Others applauded the success of their units in adapting

---

[11] Headquarters, U.S. Department of the Army (2006b, p. 2-1). Emphasis added.

to the Iraqi theater. Comments from different discussion groups included the following:

> We were always adjusting. We made adjustments to the way the METL tasks were done because the mission was always adjusting. . . . The S-2 shop [intelligence] would say "The enemy's got a different way of doing this now," so we'd adjust.

> We did the METL every day, but we adapted and evolved.

> We got the job done [the] best we could with everybody being safe.

Because unit doctrine did not always reflect or predict accurately either the activities of units or, by necessity, the way they performed standard activities, another consideration must be introduced. The DoD policy precludes the assignment of women to units whose *primary* purpose is direct combat. This implies that the doctrine of a unit must evolve to reflect a primary purpose of direct combat before a unit is closed to women. On the other hand, the Army policy precludes assignment to those units with a "routine" mission of direct combat. This difference between *primary* and *routine* is important. *Routine* is more restrictive: A support unit might develop a routine mission of self-defense while still maintaining a primary support mission. Determining or predicting a unit's routine missions is also more complicated than determining its primary mission because routine activities adapt as environmental and operational circumstances change. In other words, units can adopt new tasks that they routinely perform, but engaging in direct combat does not become the primary purpose of those units until or unless the doctrine evolves to establish this. Thus, as the routine activities of units evolve and as doctrine formally evolves, positions and units may need to be reevaluated, given the continued existence of an assignment policy for women. This will especially be the case if policymakers determine that "repelling the enemy's assault" refers to some self-defense activities in which support units currently engage. Indeed, it is worth noting that precluding the assignment of women to units that have adopted a routine mission of self-defense would likely close to women a large proportion of the deployed support units and even potentially *all units*, based on the TTPs for the FSB.

In summary, support unit personnel were trained, prepared, and expected to defend themselves and their fellow personnel; women participated in private security teams (PSTs) for military leadership, they provided security for the FOB, and they took their turns at the gates and in the guard towers. Some of these individual tasks, such as guard tower duty, could be considered ad hoc collective security responsibilities that could be assigned to individuals without violating the assignment policy. However, it was evident from the comments of some participants that self-defense tasks are routine tasks for support units. Further, there is considerable evidence from our discussions and interviews that women were trained and expected to engage in self-defense as necessary, which is consistent with the 2006 Army Posture Statement, and that the weapons they operated included crew-served weapons. If these self-defense activities equate to "repelling the enemy's assault," then including women in units whose mission routinely includes such self-defense activities is not consistent with the Army policy. This strict interpretation could have extremely significant implications for the Army, as a large number of support units might be closed to women if such self-defense activities were proscribed. Such assignments, however, would still not violate the DoD assignment policy, which neither includes repelling the enemy's assault in the definition of direct combat nor closes units unless the *primary* mission is direct combat.

Are Women Assigned to Units That Are Collocated with Direct Combat Units?

Given the uncertainty about the use of the word *collocation* in the policy, this section considers the extent to which support units and maneuver units are colocated with, or in proximity to, maneuver units, as well as the extent to which they are collocated with, or proximate to and interdependent on, maneuver units.

Are Women Assigned to Units That Are Collocated with (in Proximity to) Maneuver Units?

The Army assignment policy precludes women from being assigned to units that collocate routinely with direct combat units. As discussed earlier, there is some uncertainty about the meaning of *collocate*. If it is defined as close proximity (*colocated*), then, as one interviewee indicated, it is possible that the Army is in violation of the policy even during peacetime at home installations, since there is a mix of both support and maneuver units at many U.S. installations. For this discussion, we will assume that the policy refers only to colocation in the theater of war.

In Iraq, it is clearly evident that support units and maneuver units are colocated, or living on the same FOBs and on the same patrol bases, which, in some instances, were considerably more remote than FOBs. For example, in one instance described to us, female support personnel colocated with special forces for eight months. Another example was provided by a commander discussing a relatively remote patrol base:

> We'd send seven soldiers at a time to [patrol base name] for force protection, as perimeter guards. That included women. There were also female medics there on the medical teams . . . and there were female signaleers, who provided general support to the brigade to provide connectivity. So there were women at [the patrol base] but they didn't leave the wire unless they were going to another FOB.

FSCs, in particular, were reported to be both in very close proximity to and to have a closer de facto command relationship with the maneuver unit to which they were attached than with their assigned chain of command (the BSB). FSC personnel indicated that the commander of the maneuver unit fed, housed, and professionally evaluated the officers in the FSC, and that only paperwork and legal proceedings (the UCMJ) were transmitted between the FSC and the BSB. This paperwork and legal chain of command result in a perception among some service members that the reason FSCs are "attached" but not "assigned" to the maneuver units is to be consistent with the assignment policy. The following comments from two different focus groups reflect the perceived problems with this relationship, as well as the lack of proximity between the FSCs and their BSB chains of command. First, as noted by an FSC commander (participant 1) and first sergeant (participant 2),

> Participant 1: The paradigm we're working through is [that] the FSC commander has two bosses. [The] maneuver battalion commander rates you but [the] BSB commander is [the] senior logistician and mentors your soldiers. [This results in] conflicting relationships and expectations from both battalion staffs. The BSB commander only sees you for UCMJ actions on your soldiers. It's the worst of both worlds.
>
> Participant 2: The BSB leadership doesn't know the FSC soldiers like the maneuver guys do. It's a bad situation.
>
> Participant 1: It's hard. You see the BSB commander when you show up and say, "This soldier's a dirtbag," and the BSB commander says, "Okay, get rid of him," but I get rated by the maneuver battalion commander.

The following comments were made in another focus group of FSC commanders and first sergeants:

> Participant 1: Because of females, we're assigned to the BSB on paper, but when deployed, all FSCs are with the infantry battalion. You don't see the BSB until UCMJ. If you have issues, you have to wait to get a convoy to go to the main FOB. I was two hours away from the BSB.
>
> Participant 2: The BSB was 20 miles down the road.

In summary, women were in support units that were clearly both interacting directly with and in close proximity to maneuver units. FSCs, especially, were colocated with maneuver units, as were other support forces that included women, even on relatively remote patrol bases. In some instances, women were sent apart from the rest of their support unit to colocate and/or interact with maneuver units. Although this is not a violation of the DoD policy, it is a potential violation of the Army policy because the Army policy includes a proximity-based definition of colocation.

Are Women Assigned to Units That Are Collocated with (Proximate to and Interdependent on) Maneuver Units?

The prior discussion asserted that support units are colocated, or in proximity to, maneuver units. This section considers whether, given the alternative definition of *collocation* provided by some of our interviewees, support units are intermingled with, or proximate to and interdependent on, maneuver units. In this regard, we found mixed evidence indicating whether support units were collocated with maneuver units, and these examples relate to many issues already discussed. First, we encountered the argument that FSCs were not collocated with the maneuver battalions because the maneuver battalions could perform their missions without them (even if only for just a few days). Second, we heard that, to the extent that support units depended on maneuver units to provide security, such as convoy security, one might argue that they were collocated—that the support units could not carry out their missions, if there was greater than a level I threat, without the security provided by the maneuver units. Additionally, we were told by a commander that the immediate proximity and support of the FSCs were key to the success of his maneuver unit. His comments indicated that lives were saved because of his dependence on the FSC. Thus, if the Army policy precluding collocation might be defined as a proscription against proximate interdependence of maneuver units and units including women, it is unclear whether Army operations in Iraq are violating this aspect of the Army policy. This is not relevant to the DoD assignment policy.

Summary

This chapter considered whether the Army is currently complying with its assignment policy, given how units are operating in Iraq, as portrayed by comments in the interviews and focus groups conducted with recently returned service members. These considerations were framed around three questions: Are women assigned to maneuver units, which are the traditional direct combat units, below the brigade level? Have support units gained a direct combat mission? Are women in units that are collocated with direct combat units, defined in terms of physical proximity or proximate interdependence? This evaluation found that the Army is in compliance with the DoD assignment policy, although current assignment practices may not comply with the separate Army assignment policy.

Women are not assigned to maneuver units, and their assignments meet the "letter" of the assignment policy, but the assignments may involve activities or interactions that framers of the policy sought to rule out and that today's policymakers may or may not still want to rule out.

We found no evidence that women are initiating direct combat missions, but we found that they do interact with forces that do so, including waiting in vehicles outside. Although support units are not initiating direct combat missions, or "closing with the enemy," support units do have a self-defense mission. *Direct combat* as defined in the Army policy includes "repelling the enemy's assault." One interpretation of *repelling the enemy's assault* is that this refers only to a maneuver unit's mission to, for example, maintain control of territory in the face of an enemy onslaught. Another interpretation is that *repelling the enemy's assault* would encompass self-defense missions, even perhaps level I (individual and small-group) self-defense. Under the latter interpretation, the assignment of women to many, if not all, support units that conduct a routine self-defense mission would not be in compliance with the Army policy. Thus, whether women are participating in direct combat missions depends on the definition of *direct combat*, which differs in the DoD and Army policies, and in the Army policy will depend on the definition of *repelling the enemy's assault*.

It is also important to determine whether units are, by doctrine, conducting a primary direct combat (potentially including self-defense) mission or routine self-defense missions, or whether their current activities represent an ad hoc adjustment to the demands of the Iraqi theater. These distinctions are important because the DoD policy would preclude the assignment of women to those units with a *primary* direct combat mission (defined by the DoD policy), whereas the Army policy would prevent the assignment of women to those units with even a *routine* direct combat mission (defined by the Army policy). Support units conduct routine self-defense missions, and some support units appear to provide security for other support units, but it is not clear whether the latter is routine or ad hoc. If any of these missions are included in the definition of *repelling the enemy's assault* and the proscribed activities are routine, then the assignment of women to those units would be inconsistent with Army policy. However, assignment to these units is consistent with DoD policy, which focuses on the *primary* mission of the unit and which does not include *repelling the enemy's assault* in the direct combat definition.

The DoD policy indicates that the military services can include certain restrictions, such as those on collocation, and the Army policy does preclude women from being assigned to units that collocate with direct combat units and subsequently defines *collocation* in terms of proximity. We find considerable evidence that support units are in proximity to direct combat units. However, some experts have argued that collocation is actually defined to include proximity and interdependency. Although that argument appears to be unsupported by the definition provided in the Army policy, we also explore the implications of that definition. Given that definition, the evidence is inconclusive. While some might maintain that the ability of maneuver units to accomplish their missions independently, even for a limited number of days, disproves collocation, others could argue that the maneuver units' dependency on FSCs, or even the supply units' dependency on maneuver units for security, does constitute collocation. Neither proximity nor proximate interdependence between support and maneuver units would violate the DoD assignment policy, but the assignment of women to support units in Iraq may not be consistent with the Army's assignment policy, depending on the definition of collocation.

CHAPTER FOUR

Is the Assignment Policy Appropriate for Future Military Operations?

Chapter Three focused on whether the Army is currently complying with the assignment policies for Army women, given how units are operating in Iraq. It did not challenge whether the language and concepts in the policy were appropriate or could be improved. This chapter considers a different question and explores the DoD and Army assignment policies themselves, in the context of the Army's transformation and operations in Iraq, to assess whether the language and concepts of the Army and DoD assignment policies are appropriate for military operations in the future, which could resemble Army operations in Iraq, and to the Army's new force structure.

The Army is currently emphasizing the development of leaders who are "multi-skilled, innovative, agile, and versatile."[1] This adaptability and versatility of personnel is important not only because of the conditions in Iraq but also because of how doctrine for the modular BCTs is evolving. Yet the question remains: Does the DoD or Army assignment policy constrain the Army's flexibility and responsiveness and, consequently, its effectiveness in Iraq—and, quite easily, other operations as well—beyond what leaders and policymakers might find acceptable?

This chapter assesses whether the language and concepts in the DoD and Army assignment policies are appropriate for the new military environment and the Army's new structure. We address this by

---

[1] U.S. Department of the Army (2006, p. 15).

specifically considering Army operations in Iraq. We explore six aspects of the assignment policies, including

- the restrictions on assigning women to direct combat units
- the focus on a linear battlefield and engaging and closing with an identifiable enemy force
- the restrictions on collocation with direct combat units
- the restriction on repelling the enemy's assault
- the identification of occupations and units that are open or closed to women
- the focus on the assignment, rather than the tasking or employment, of women.

For each of these aspects, we consider what might be the intent of policymakers as a way of determining the appropriateness of the language in the assignment policy for future military operations, given that such operations may resemble operations in Iraq, and for the Army force structure. The discussion is intended to assist policymakers as they determine the sufficiency of the current assignment policy and, if necessary, the content of a revised assignment policy.

As a point of departure, we begin with broad assessments and comments by returned service members about the compatibility of the Army assignment policy to the Iraqi theater and to Army operations in Iraq. The comments provide background information about how the assignment policy relates to actual operations in Iraq, as perceived by personnel who were there. The comments indicate that Army personnel believe that there is tension between the current situation in Iraq with respect to Army women and adherence to the 1992 Army assignment policy. Given a choice, most of the Army personnel who participated in our study were inclined to resolve that tension by revising or revoking the policy. It is likely that many of these perceptions were based on very strict interpretations (or misinterpretations) of the assignment policy, and thus a revision that clarifies the terminology and definitions of the policy might possibly address and resolve many of these negative perceptions. Although these observations are in the context of current, rather than future, operations, they are important in that at least some future operations may share characteristics with the current operations in Iraq.

Perceptions of Army Personnel Who Served in Iraq: How Compatible Is the Assignment Policy with the Iraqi Theater?

During our interviews and focus groups, we read the following section of the Army's assignment policy and asked participants to discuss its relevance to operations in Iraq:

> The Army's assignment policy for female soldiers allows women to serve in any officer or enlisted specialty or position except in those specialties, positions, or units (battalion size or smaller) which are assigned a routine mission to engage in direct combat, or which collocate routinely with units assigned a direct combat mission.[2]

Personnel who were already familiar with the policy, as well as those who were not, asserted that the policy did not reflect the environment in Iraq. Some of those less familiar with the policy even expressed surprise or disbelief at its content, as indicated by comments such as, "I had no idea," or laughter during the discussion of the above excerpt and the definitions of *direct combat* and *collocate* that accompanied it.

Recently returned Army personnel described the policy as "very obsolete," "archaic," and "a step back." Interview and focus group participants also focused on the lack of meaning in the words, as discussed in Chapter Two, and their perception of the inappropriateness of the policy constraints to the operations they had experienced:

> If women are on a convoy, they are subject to direct combat with the enemy. You were primarily subject to attack every time you went out there. The paradigm of direct fire and engagement is just no longer there. It's a different environment than when the policy

---

2   Headquarters, U.S. Department of the Army (1992, p. 1).

was written in 1992. Any solider in the theater—if they leave one FOB to go to another—they are subject to direct engagement.

Whoa. They need to rewrite the policy on women. This war is a different species. Every time [women are] outside the FOB on convoy escort, they're in the midst of anything that would come up, and they had to react, fire downrange. If you hit IEDs, they had to react.

Some Army personnel expressly addressed whether the Army was compelled to violate its own policy with respect to the assignment of women in Iraq. For certain operations, it was not clear to Army personnel whether the Army was adhering to the policy or not. As one commander stated,

There's a fine line between violating DoD policy and assigning women to gun trucks. The gun trucks are not intended for front-line combat, but any commander knows that, in reality, that could happen. If the intent is to prevent women from experiencing combat, we're past that.

But others did not feel that preventing women from experiencing combat was the intent of the policy. Indeed, those familiar with the Army's policy tended to believe that it was referring to offensive maneuvers intended to close deliberately with the enemy; repelling an enemy's assault while on a convoy or FOB was not always seen as direct combat. As one individual put it, "We try to follow whatever DoD puts out. They don't want females intentionally on raids, killing and kicking in doors." Many of the comments can be summarized by that of one commander: "Those officers that understand the policy would say it's not relevant. But most would not understand it." In other words, it is quite possible that many of those who characterized the policy as "archaic" or "obsolete" are interpreting it more restrictively than it was intended to be interpreted. This underscores the need to resolve the ambiguity of the policy. At a minimum, the policy and its intent beg clarification.

Perceptions of Army Personnel Who Served in Iraq: Is the Current Policy Conducive to Today's Army and to Army Operations in Iraq?

Despite their disparate interpretations of the assignment policy, Army personnel were consistent in their perception that a strict adherence to the Army policy would have negative implications. One officer simply stated, "It [the policy] doesn't make sense in today's Army," while others gave more specific examples of how the policy could be detrimental to both Army personnel and operations. The policy was regarded as a type of discrimination that would eliminate job opportunities, including leadership opportunities, for female personnel. Additionally, the policy had a perceived downside for male personnel, not only because they would feel that they would have to compensate for what the women would not be permitted to do if the policy were strictly interpreted, but also because they believed that policy seemed to place a higher value on women's safety. One female soldier remarked, "Why is a woman's life worth more than a man's? I think a soldier's a soldier," and a male officer stated, "I'm sick of reading articles where they say mothers shouldn't be in this position."

Absolute compliance with the Army assignment policy, if strictly interpreted, also could result in negative consequences for Army operations. At a very basic level, one individual mentioned that one commander's attempt to enforce his interpretation of the policy by isolating the women from their male peers not only was poorly received by the women in the unit but also undermined the platoon leader's and platoon sergeant's abilities to communicate effectively with their soldiers.

Additionally, it was widely accepted that female service members were necessary to Army operations in Iraq and that they were needed to interact with female Iraqi civilians during searches, raids, and the operation of checkpoints. The guidance on the conduct of these operations is relatively clear. According to Army Field Manual 3-20.96, *Heavy Brigade Combat Team Reconnaissance Squadron*, when checkpoints are used in stability and support operations, female service members should be used to search females. It notes, "Use female searchers. If female searchers are not available, use doctors, medics, or members of the local populace."[3] Importantly, the heavy brigade combat team (HBCT) reconnaissance squadron does not have any positions coded for women.[4] This means that, to fulfill this requirement, women must be taken from somewhere else, perhaps an FSC or some other support unit.

This is consistent with comments from returned service members, such as the following:

> There is a female search team requirement. The [unit name] tells us, "You need to resource all female search teams." That depleted our unit a lot. All those female search teams. Plus the women that were needed at the entry points of the FOB. Plus the women needed in the dining facility to search, and also those needed at the third-country national interviews. There's only a finite number of women in a unit. They [mission commanders] didn't consider the MOS that they would hinder.

The importance given to respecting the Iraqi culture also meant that female medics were sought out on a regular basis to interact with Iraqi women in their capacity as health care providers. Other examples of tasks for which women were needed or requested included processing and interrogating female detainees and simply interacting or talking with Iraqi women when a team went outside the wire to conduct a mission.

While the need to include women on search teams was often a requirement, unit leaders were well aware that using women in the aforementioned capacities offered clear strategic advantages, as the following comments demonstrate:

> I think females should be in combat units to connect with the female population in Iraq. We're missing lots of intelligence. Males can't talk to Iraqi females.

---

[3]   Headquarters, U.S. Department of the Army, *Heavy Brigade Combat Team Reconnaissance Squadron*, Washington, D.C., FM 3-20.96, March 15, 2005a, para. 8-135.

[4]   Women are excluded from the reconnaissance squadron because it has a primary mission of direct combat.

Sometimes we got better information from the wives than from the Iraqi men. We'd send medics to talk to the females.

I had two female interpreters that worked for me. When I had detainees with mental illness, I tried to use a female interpreter. The mothering effect could cut down on [their] anxiety level.

The officer who made the last comment also stated there could be cultural disadvantages to using female soldiers. Specifically, he indicated that some male detainees would not listen to orders given by female guards and would not tell female medics what was ailing them. However, these did not appear to be insurmountable obstacles; as the officer put it, "Detainees' social behavior affected how we used females, but we learned how to use them." The cultural advantages to using women appeared to exceed any cultural disadvantages in this context.

Much more broadly, the policy was also regarded by some as a constraint to Army modularization:[5]

The assignment policy is what's holding us back from truly transforming. The sticking point about whom they [the FSCs] work for is women. If we're truly transforming, the FSCs would be assigned to maneuver battalions. We've bastardized the modular concept. Everything we do, except for the rating, says they belong to the BSB. To Big Army, they [the FSCs] are 100 percent [BSB]. But in Iraq, it doesn't feel that way and it doesn't need to be that way, either. It would probably be better if [FSCs were] just assigned to maneuver battalions.

Perhaps most important, officers believed not only that accomplishing their missions would be a challenge without women in their current capacities, but also that strict total adherence to the policy meant that women would have a much more limited presence in Iraq,

---

[5] There are other possible reasons for having the FSCs attached, as opposed to assigned, to a maneuver battalion. These include efficiencies in combat service support (CSS) training across a division and the professional development (i.e., the mentorship) of FSC commanders. However, it is not our intent to contrast the advantages of how the FSC is organized, but instead to record the sentiment of those interviewed about the assignment policy.

which would considerably constrain the resources available to commanders. For instance, they felt that to comply with the policy, women in the medical field might need to be restricted to the hospitals in Iraq, while other individuals went even further with comments such as, "There should be no women in Iraq, based on those criteria [in the assignment policy]," and "We would have to keep them [women] in the U.S." While others would maintain that this is a misinterpretation of the policy, the implications are clear: Some very strict interpretations (or misinterpretations) of the policy could preclude women from deploying to a combat theater such as Iraq. Not surprisingly, then, some personnel claimed, as noted earlier in this chapter and in the comments that follow, there simply were not enough personnel to do the job without women:

> We need them, the female logisticians. It's America. Can't accomplish the mission without women.

> Reality is we can't fill FSCs without females.

> The way I see it, there are two choices. One, comply to the spirit of the law and not have our trucks work [because women are needed for this], or two, keep them [women] on the BSB books and we'll just have to live with them. That way they live with the troops and fix our stuff.

Our discussions with returned service members also addressed the extent to which they perceived that tasks were being assigned by gender or believed that there were performance differences between men and women. In general, although some service members perceived that men and women were being assigned different tasks, most reported that task assignments were gender-neutral, with the notable exception of those tasks, discussed previously, that required women. Likewise, service members tended not to see performance differences by gender, and they also acknowledged that some tasks, such as interacting with civilian Iraqi women, required female service members. Thus, the reality that unit positions likely could not be filled without women, along with the perceived advantages of having women available to interact with female Iraqis in various capacities, the belief that there were no systematic performance differences between men and women, and the general preference to assign tasks in a gender-neutral way, all suggest that the Army personnel we spoke with tended to support the current use of women in the Iraq environment. Further, some individuals even proposed expanding their involvement. However, comments from our interviews and focus groups also indicate that Army personnel believe that there is tension between the current situation in Iraq with respect to Army women and adherence to the 1992 Army assignment policy. Given a choice, most of the Army personnel who participated in our study were inclined to resolve that tension by revising or revoking the policy. It is likely that many of these perceptions were based on very strict interpretations (or misinterpretations) of the assignment policy, and thus a revision that clarifies the terminology and definitions of the policy might possibly address and resolve many of these negative perceptions.

The Appropriateness of Different Aspects of the Assignment Policy

This section assesses whether the language and concepts in the DoD and Army assignment policies are appropriate for the new military environment and the Army's new structure by exploring the six aspects of the assignment policies listed in the beginning of this chapter. For each of these aspects, the discussion considers what might be the intent of policymakers as a way of determining the appropriateness of the language in the assignment policy to future military operations, given that they may resemble operations in Iraq, and to Army force structure. This discussion is intended to assist policymakers as they determine the sufficiency of the current assignment policy and, if necessary, the content of a revised assignment policy.

The Restrictions on Assigning Women to Direct Combat Units

Both the DoD and the Army assignment policies preclude the assignment of women to direct combat units, though they differ in terms of whether those units must have a primary mission, or merely a routine mission, of direct combat. In light of the Army's modularization and the way in which maneuver and support units interact in Iraq, this language is appropriate if the intent of the policy is to prevent women from initiating direct combat or becoming soldiers whose primary purpose is to engage and destroy the enemy. This objective reflects the concern of those who do not want women to become "steely-eyed killers."

This restriction is also appropriate if the intent of the policy is to ensure that the official chain of command for support units is through other support units and that, for example, the in-garrison oversight and the professional mentoring of support officers is provided by other support officers through their own chain of command, such as the BSB commander overseeing and developing the FSC officers.

This restriction is neither appropriate nor sufficient if the intent is to keep women physically closer to their own chains of command, since we heard from FSC personnel about instances in which they were collocated with the maneuver units but were several hours away from the BSB command.

This restriction is neither appropriate nor sufficient if the intent is to preclude close and regular interaction between the women in support units and the combat arms units, both inside and outside the safety of the installation, or to preclude women from directly supporting direct combat missions, such as in the case of HUMINT or psychological operations (PSYOP) personnel or in the instance of female medics accompanying combat arms personnel on missions off the installation.

And finally, this restriction is not appropriate or sufficient if the intent is to prevent women from participating in all forms of direct combat, including self-defense (should "repelling the enemy's assault" be determined to include self-defense).

The Focus on a Linear Battlefield Context and Engaging and Closing with an Identifiable Enemy Force in the Definition of *Direct Combat*

The DoD policy specifies that direct ground combat takes place "well forward on the battlefield," and both the DoD and the Army policies depend on the identification of an enemy to define *direct combat* and thus to identify those units to which women may not be assigned. This monograph has already discussed the difficulty of defining these terms in the context of Iraq.

There are also other reasons that this reliance on a linear battlefield and a defined enemy is inappropriate. If the intent of the emphasis on precluding women from being far forward is rooted in concern about keeping women safe from either bodily harm or from capture, no one in Iraq is safe. Service members generally agreed that anyone who ventures off the installation is vulnerable to the enemy, and most felt that even the FOBs were not safe, given that they were vulnerable to mortar attacks.

This definition of ground combat is not appropriate if the intent is to deploy women to the theater but to preclude them from all engagements with the enemy, including self-defense. As stated previously, all deployed Army units are trained and prepared for the self-defense mission, and this is a practice that predates the Iraq war and the GWOT.

This distinction is also no longer effective to preclude women from participating in close and regular interaction with combat arms units, both inside and outside the safety of the installation.

Finally, this aspect of the policy does not keep women from interacting with the enemy, if that was the intent. Indeed, Army doctrine requires female service members to interact directly with civilian Iraqi women, who might be the enemy.

The Restrictions on Collocation with Direct Combat Units

The assignment policies state that women cannot be assigned to units that collocate with a direct combat unit. This aspect of the policy needs to be clarified. If the intent is to prevent the deployment of women in proximity to direct combat units, this aspect of the policy is neither appropriate nor feasible, given operations and installations in Iraq. Given that the Army restriction on collocation does not explicitly pertain only to combat theaters, if policymakers did confirm that the intent is to ban support units from being in immediate proximity to maneuver units, then they would need to determine whether this ban applied to stateside installations as well.

If the intent is to preclude close and regular mission-related inter-action of women in support units with combat arms units, then a well-defined ban on collocation would be sufficient. However, such a ban would limit the career opportunities of military women in certain occupations (such as those in FSCs), would limit the Army's assignment flexibility, and would likely limit the Army's ability to fill FSC and other billets.

Further, if the objective is to limit the interaction between support units with women and direct combat units, then a ban on collocation would be appropriate if policymakers did not want women either exposed to the direct combat actions of maneuver units or protected by maneuver units. In other words, such a ban could increase the risk to women in support units that currently benefit from the additional security provided by maneuver units.

Finally, such a ban would be appropriate only if policymakers either intended to exclude women from jobs they successfully performed in Iraq or accepted that this change would do so.

The Restriction on Repelling the Enemy's Assault

The Army assignment policy defines direct combat as either closing with the enemy or repelling its assault. Repelling an assault is sometimes interpreted as self-defense. Interviews and focus groups with returned personnel indicated that support units are actively involved in their own self-defense and that some support units are providing self-defense for other support units. If the intent of this clause is to proscribe individual and group self-defense, it is inconsistent with and inappropriate for the evolved battlefield, which lacks a safe rear area. If this is not the intent, the clause needs to be clarified. Thus, this restriction, which is found in the Army policy but not the DoD policy, should be removed or clarified.

The Identification of Occupations and Units That Are Open and Closed to Women

The Army assignment policy specifies that women can serve in any specialty, position, or unit, except those that have a routine direct combat mission or are collocated with direct combat units. Thus, entire units are closed to women, although there are support jobs within those units that are very similar to other jobs that women are filling (such as supply clerks and cooks). If the intent is a focus on the immediate coworkers and chain of command, and if the intent is to preclude women from interacting with combat arms personnel at this level, the emphasis on units is appropriate.

This emphasis on closing entire units is also appropriate if there is concern that all members of a unit may need to perform the primary mission of offensive ground combat, that they may all need to become infantrymen, for example, and participate in offensive missions.

This emphasis on closing entire units is insufficient if the objective is to preclude women from participating in any engagement with the enemy, as women are currently participating in self-defense missions despite the closure of specified units. This emphasis is also insufficient if the intent is to preclude women from being tasked to participate in the work of closed units. For example, combat medics are all male, but one officer reported that three-fourths of the medics he was sending to supplement the combat medic platoon were female. This practice complies with the current policy because these women are not assigned to be combat medics; they are simply tasked to relieve combat medics.

The Focus on the Assignment of Women Rather Than the Employment of Women

As mentioned in Chapter One and discussed in more depth in Appendix B, both the DoD and Army assignment policies constrain the units to which women can be assigned. The Army assignment policy also states,

> Once properly assigned, female soldiers are subject to the same utilization policies as their male counterparts. In event of hostilities, female soldiers will remain with their assigned units and continue to perform their assigned duties.[6]

---

[6] Headquarters, U.S. Department of the Army (1992, p. 2).

As in the case of female medics, accounted earlier, commanders can use female service members to respond to operational needs within the theater by assigning them tasks similar to those assigned to their male counterparts. Commanders are also not constrained from using any of their personnel for non-METL tasks or for METL tasks conducted differently from what the doctrine might otherwise suggest. This distinction between the assignment of women and their employment is inappropriate if policymakers object to women taking on roles or serving with units to which they cannot be assigned.

This distinction is appropriate if a desired objective is to give priority to military effectiveness and, thereby, provide the commander with the best opportunity to respond to a changing environment and to accomplish the unit's mission most effectively and safely with the best people at hand. To limit the ways in which commanders could use their personnel would likely limit their flexibility and military effectiveness.

Summary

This chapter considered whether the current assignment policy is appropriate for the new military environment and the Army's new structure. Our focus was on the language and the concepts in the policy, to assess whether they are appropriate for military operations in the future, which could resemble Army operations in Iraq, and for the Army's new force structure. As a point of departure, we considered the perceptions of returned service members regarding the current assignment policy and Army operations in Iraq. We found that many returned service members were not aware of the assignment policy or were not familiar with the specific details of the policy. This is not surprising, given that adhering to the assignment policy is generally not the responsibility of tactical commanders in the theater, as they are not assigning individuals to units. Those returned service members who were aware of the policy often did not understand it, and some found it generally inapplicable to the Iraqi theater. Some personnel also expressed opinions that the policy was a backward step from the successful execution of the mission in Iraq, in which women have been involved in many aspects of operations.

If the assignment policy is intended to preclude support units from having self-defense missions and capabilities, the focus on a defined enemy and the linear battlefield context is inappropriate for the Iraqi theater, as is the Army restriction against women in units with a mission to repel the enemy's assault. Thus, the latter clause should be either removed or clarified. The appropriateness of other aspects, such as the restrictions on assigning women to direct combat units, the restrictions on collocation with direct combat units, and the identification of both occupations and units that are open or closed to women, is a matter of interpretation and judgment and depends on the objectives, or intent, of the policymakers. For example, the restriction against assigning women to maneuver units does keep women from being part of a unit that initiates direct combat, or closes with the enemy. However, none of these restrictions precludes women from interacting closely with maneuver unit personnel or from interacting with the enemy or with potential enemy personnel. These restrictions do ensure that support units (and the women in them) are trained and mentored by other support unit personnel while in garrison, but they do not ensure closer proximity to the support unit chain of command than to maneuver units while in the theater. Finally, these restrictions, if strictly enforced, could exclude support units from the benefit of extra security provided by maneuver units and could eliminate female service members from jobs they have performed successfully in Iraq.

Finally, the focus in the policy on assignment to units, rather than the individual employment of women, is especially important, and the appropriateness of this aspect depends on whether military effectiveness and flexibility are determined by policymakers to be more important than prescribing the precise activities in which women can engage.

CHAPTER FIVE

Conclusions and Recommendations

Conclusions

This monograph has considered whether the current DoD and Army assignment policies for women are understandable, whether the Army is complying with the DoD and the Army policies, and whether the language and concepts in the assignment policies are appropriate for future operations, given what we have learned from Army operations in Iraq.

We find that the precise meaning, or "letter" of each policy is not clearly understood, largely because of the asymmetric nature of warfare and the nonlinear battlefield in Iraq, which renders important elements of the policy, such as the terms *forward* and *enemy* less meaningful. Additionally, the meaning of the term *collocate* is not clear. Nor is the "spirit" of each policy clearly discernible, since there is not agreement about whether the policy is designed to protect women (either from bodily harm or from the risk of capture) or about why or to what extent women should be kept from direct combat. While DoD personnel were more likely to emphasize objectives such as maximizing operational effectiveness and maximizing assignment flexibility over objectives to protect male or female service members, the members of Congress who were interviewed were less likely to be in agreement.

This research also assessed whether the Army is complying with the DoD or Army assignment policy. This involved addressing three questions: Are women assigned to direct combat units, i.e., maneuver units, battalion size or smaller? Have the support units to which women are assigned gained a mission of direct combat? Are women in units that are collocated with direct combat units? This research found that the Army is complying with DoD policy but may not be in compliance with its own policy.

This research found that women are not assigned to direct combat units. For example, FSCs have a direct support relationship with a maneuver unit and are often under the operational control of the unit they support. However, particularly in the case of FSCs, some personnel characterized this command relationship with maneuver units (they were attached, not assigned) as only a semantic distinction because they are fed, housed, and led by the maneuver units.[1] Although these assignments meet the "letter" of the assignment policy, the assignments may involve activities or interactions that framers of the policy sought to rule out and that today's policymakers may or may not still want to rule out.

Support units in Iraq do participate in routine self-defense missions and may also provide security for other units. Thus, women are assigned to units with a routine mission of self-defense. This is consistent with the DoD assignment policy, because the primary mission of a support unit is not direct combat on the ground, and DoD policy states that women shall be "excluded from assignment to units below the brigade level whose primary mission is to engage in direct combat on the ground." However, support units routinely engage in self-defense missions. If such missions are included in the definition of *repelling the enemy's assault* and thereby interpreted as a routine direct combat mission, then the Army is out of compliance with its assignment policy. Recall that Army policy stipulates excluding women from "specialties, positions, or units (battalion size or smaller) which are assigned a routine mission to engage in direct combat." But if such missions are not interpreted as direct combat missions, then the Army is not out of compliance. In any event, it is germane to this debate that women in

---

[1]   Assigning a unit places the complete control of the unit under the command to which it is assigned. When these units were attached to the maneuver units, the maneuver units often had operational control over and, generally, overall responsibility for and authority over the support unit, except for UCMJ responsibilities.

support units in Iraq do routinely participate in individual and group self-defense missions.

In sum, support units and support personnel are currently performing routine missions, such as self-defense, that might, according to a very strict interpretation of the Army policy, have features of direct combat.[2] If these missions were classified as direct combat, women would be precluded from assignment to those units. It becomes important to confirm both whether support units do indeed have a routine mission deemed to be direct combat or, alternatively, whether the phrase *repelling the enemy's assault by fire, close combat, or counterattack* refers only to the mission of maneuver units (not to the mission of support units). This interpretation appears to be most consistent with the wording in maneuver unit doctrine. If routine self-defense missions were proscribed by this policy, there would be severe implications for the assignment of women, as many, if not all, support units could be closed to women. It is, however, important to note that assignment to units with a *routine* mission of self-defense (or even other direct combat) does not violate the DoD assignment policy, which only constrains assignment to those units with a *primary* mission of direct combat and does not include repelling the enemy's assault in its definition of direct combat.

The Army policy also stipulates the exclusion of women from units that collocate with direct combat units. Whether or not women are in units that collocate with maneuver units depends on the definition of *collocation*. Women are in units that are colocating with (or working in proximity to) direct combat units, since both support units and maneuver units reside together on the FOB. Women are also leaving the FOB with maneuver units, in support of either a combat tasking or an operation, or with maneuver units providing additional security for

---

2   Again, the Army definition of *direct combat* is

> Engaging an enemy with individual or crew served weapons while being exposed to direct enemy fire, a high probability of direct physical contact with the enemy's personnel and a substantial risk of capture. Direct combat takes place while closing with the enemy by fire, maneuver, and shock effect in order to destroy or capture the enemy, *or while repelling the enemy's assault by fire, close combat, or counterattack.* (Headquarters, U.S. Department of the Army, 1992, p. 5; emphasis added)

the support mission, such as during convoys. Whether that interaction outside the FOB equates to a proximate interdependency, or collocation, of support and maneuver units is unclear.

This monograph also evaluated whether the concepts and language in the current policy for assigning women were appropriate for future military operations, given the Army's experience in Iraq. The attitudes and perceptions of returned service members regarding the assignment policy provide a useful context for this evaluation. Many personnel recently returned from Iraq did not know about the policy, as they were not generally involved in the assignment of personnel to units. Those who were familiar with the assignment policy did not generally find it understandable or useful. Some felt that it was a backward step from operations that were being conducted successfully in Iraq. Although many of their perceptions may be based on misinterpretations of the policy, their attitudes confirm both the confusion about the current assignment policy, as discussed in Chapter Two, as well as the relevance of our analysis of the appropriateness of specific wording in the assignment policy.

This research specifically considered the appropriateness of different aspects of the current DoD and Army assignment policies, including the following: the restrictions on assigning women to direct combat units; the focus on a linear battlefield and engaging and closing with an identifiable enemy force; the restrictions on collocation with direct combat units; the restriction on repelling the enemy's assault; the identification of units that are open and closed to women; and the focus on assignment, rather than the tasking or employment of women.

Of these aspects of the assignment policy, the assumption of a defined enemy in a linear battlefield context is inappropriate for future operations. The restriction against repelling the enemy's assault requires clarification. If it is determined to include self-defense missions, then this clause is also inappropriate for future operations, as it substantially (or completely) restricts the assignment of women to deploying units because support units are expected to engage routinely in self-defense missions. The appropriateness of other aspects of the assignment policy is a matter of interpretation and judgment and depends on the objectives or intent of policymakers. For example, the restriction against assigning women to maneuver units does keep women from being part of a unit that initiates direct combat, or closes with the enemy. However, neither the restriction against assigning women to direct combat units nor the collocation restriction precludes women from interacting closely with maneuver unit personnel or from interacting with the enemy or with potential enemy personnel. These restrictions do ensure that support units (and the women in them) are trained and mentored by other support unit personnel while in garrison by requiring those support units to be assigned only to support units, but they do not ensure closer proximity to the support unit chain of command than to maneuver units while in the theater. These restrictions could be interpreted to exclude support units from the benefit of the extra security of maneuver units, and could eliminate female service members from jobs they have performed successfully in Iraq. Indeed, a very strict interpretation of the Army assignment policy could preclude assigning women to almost any support unit that would deploy to a future Iraq-type conflict because almost all deploying support units could expect to perform self-defense missions routinely. If the policy is so strictly interpreted that self-defense is included in the direct combat mission, such support units would be closed to women.

Finally, the appropriateness of the current policy's focus on the assignment of women to units rather than the employment of individual women depends on whether military effectiveness and flexibility is determined by policymakers to be more important than prescribing the precise activities in which women can engage. Military effectiveness and flexibility entail adapting to changes in enemy strategy, tactics, and weapons, and this implies that commanders may need to employ military resources, including individual women and units with women, in ways not initially envisioned in policy and possibly not well covered in doctrine. The Iraq example has shown how the application of the current assignment policy has led to the employment of units that include women in ways that are consistent with the DoD assignment policy, might not be consistent with the Army assignment policy, and, yet, based on our interviews and focus groups, have been consistent with maintaining unit effectiveness and capability.

Recommendations

This research effort was established to assess the extent to which the current assignment policy is appropriate to, and reflected in, Army doctrine, transformation, and operations in Iraq. The intent of this research is not to prescribe policy, but rather to report research findings about the current assignment policy and Army operations in Iraq and to identify issues in current policy, doctrine, or employment for DoD's consideration.

The critical first issue is whether there should be an assignment policy for military women. The interviews and discussion groups were not representative of the Army, so we cannot determine the prevailing views of service members. Nonetheless, at least some personnel believe that there should not be an assignment policy, meaning simply that all positions should be gender-neutral. This is certainly not unanimous. Other service members, advocacy groups, and some voices on Capitol Hill clearly feel that there should be an assignment policy. Regardless of the content of the assignment policy, articulating the need for an assignment policy is a worthwhile first step.

If there is a continued need for an assignment policy, we recommend these considerations to guide its design, implementation, and any legal reporting requirement:

- *Recraft the assignment policy for women to make it conform—and clarify how it conforms—to the nature of warfare today and in the future, and plan to review the policy periodically.* Given the lack of common understanding about either the intent or the specifics, the current policy is not actionable. At an absolute minimum, it must be better articulated. Periodic reviews of the policy would ensure that the language reflects the evolving nature of the military mission.
- *Make clear the objectives or intent of any future policy.* Given the uncertainty of future operations and the likely inability to predict them accurately, public understanding of the spirit of the policy will best support the continued implementation of the

policy, even if the details of the policy are not appropriate to all future military operations.

- *Clarify whether and how much the assignment policy should constrain military effectiveness, and determine the extent to which military efficiency and expediency can overrule the assignment policy.* For example, should the military leadership be able to change the policy without a month or more of notice to Congress?
- *Consider whether a prospective policy should exclude women from units and positions in which they have performed successfully in Iraq.* If the current assignment policy is continued and interpreted strictly, women would likely be removed from many units, and even some of the occupations, in which they currently serve. Thus, a revised assignment policy will need to address whether any of the opportunities currently open to women, or the positions in which they currently serve, can be closed. The 1994 Aspin memo, for example, specified that the guidance would be used "to expand opportunities for women. No units or positions previously open to women will be closed under these instructions."
- *Given that the assignment policy is unusual because of the legal requirement to report policy changes to Congress, consider the extent to which an individual service policy should differ from overall DoD policy.* If individual service policies remain separate from DoD policy, and if there remains a legal requirement to inform Congress of changes to the policies or of divergence from the policy, recognize that these differing policies could present reporting challenges.
- *Determine whether an assignment policy should restrict women from specified occupations or from both occupations and units.* This is related, in part, to the issue of proximity. For example, should women be precluded from interacting with personnel in direct combat occupations, or should they be precluded from certain occupations, such as infantry? The practical implication of the latter is that, if women are excluded from select occupations, there might be female support personnel (e.g., cooks or supply personnel) in an infantry battalion, but other jobs, such as infantry jobs in that same battalion, would remain closed to women.

Indeed, some jobs, such as teaching infantry tactics in a school-house, would remain closed to women even while the unit (in this example, the schoolhouse) was open. If proximity is determined acceptable, one benefit of closing only occupations to women is the flexibility the Army would have to make revisions, since each unit's modification table of organization and equipment would not require gender-coding review. Instead, billets would be inherently open or closed based on the occupation of that unit. However, this needs to be further examined to ensure that it does not result in the closing of entire branches (such as field artillery) that currently have many positions open to women. In addition, policymakers need to determine whether such a change would place women in positions inconsistent with a revised policy on direct combat, such as those in units (like certain engineer companies) expected to perform an infantry mission as needed.

- *Determine whether colocation (proximity) and collocation (proximity and interdependence) are objectionable, and clearly define those terms should they be used in the policy.*

- *If unit sizes (or levels of command) are specified in the assignment policy, make apparent the reason and intent for specifying unit size, given that modularization and the context of an evolving battlefield may negate this distinction.*

- *Consider whether the policy should remain focused on assignment to units rather than the employment of individual women.* An assignment policy for women does not constrain what individual women can actually do while deployed, and, thus, it provides military commanders with the maximum amount of flexibility to complete their missions most effectively and safely while making best use of their personnel resources.

Aspin 1994 Memorandum

THE SECRETARY OF DEFENSE

WASHINGTON, DC 20301-1000

January 13, 1994

MEMORANDUM FOR THE SECRETARY OF THE ARMY
                              SECRETARY OF THE NAVY
                              SECRETARY OF THE AIR FORCE
                              CHAIRMAN, JOINT CHIEFS OF STAFF
                              ASSISTANT SECRETARY OF DEFENSE
                                 (PERSONNEL AND READINESS)
                              ASSISTANT SECRETARY OF DEFENSE
                                 (RESERVE AFFAIRS)

SUBJECT: DIRECT GROUND COMBAT DEFINITION AND ASSIGNMENT RULE

References: (a) SECDEF memo, April 28, 1993
              (b) SECDEF memo, February 2, 1988
              (c) FY94 National Defense Authorization Act

My memorandum dated April 28, 1993, directed the Military Services to open more specialties and assignments to women and established an Implementation Committee to ensure that those policies are applied consistently. I also charged the Committee to review and make recommendations on several specific implementation issues.

The Committee has completed its first such review, that of the "appropriateness of the 'Risk Rule'", reference (b), and concluded that, as written, the risk rule is no longer appropriate. Accordingly, effective October 1, 1994, reference (b) is rescinded.

My memorandum restricted women from direct combat on the ground. The Committee studied this and recommended that a ground combat rule be established for assignment of women in the Armed Forces. Accordingly, the following direct ground combat assignment rule, and accompanying definition of "direct ground combat," are adopted effective October 1, 1994 and will remain in effect until further notice.

   A. Rule. Service members are eligible to be assigned to all positions for which they are qualified, except that women shall be excluded from assignment to units below the brigade level whose primary mission is to engage in direct combat on the ground, as defined below.

   B. Definition. Direct ground combat is engaging an enemy on the ground with individual or crew served weapons, while being exposed to hostile fire and to a high probability of direct physical contact with the hostile force's personnel. Direct ground combat takes place well forward on the battlefield while locating and closing with the enemy to defeat them by fire, maneuver, or shock effect.

The Services will use this guidance to expand opportunities for women. No units or positions previously open to women will be closed under these instructions.

The Services will provide the Assistant Secretary of Defense (Personnel and Readiness), not later than May 1, 1994, with lists of all units and positions closed to women and their proposed status based on implementation of this policy. These lists will be arrayed in three columns: jobs currently closed that are proposed to be opened, jobs currently closed that are proposed to remain closed, and a column justifying each entry. The proposed changes will be reviewed by the Implementation Committee and the Assistant Secretary of Defense (Personnel and Readiness). The Services will then coordinate approved implementing policies and regulations with the Assistant Secretary of Defense (Personnel and Readiness) prior to their issuance. These policies and regulations may include the following restrictions on the assignment of women:

- where the Service Secretary attests that the costs of appropriate berthing and privacy arrangements are prohibitive;

- where units and positions are doctrinally required to physically collocate and remain with direct ground combat units that are closed to women;

- where units are engaged in long range reconnaissance operations and Special Operations Forces missions; and

- where job related physical requirements would necessarily exclude the vast majority of women Service members.

The Services may propose additional exceptions, together with the justification to the Assistance Secretary of Defense (Personnel and Readiness). The process described above will enable the Department to proceed with expanding opportunities for women in the military as well as comply with the reporting requirements contained in Section 542 of reference (c).

The Difference Between an Assignment Policy and an Employment Policy

The policy regarding the roles of women in the Army is an assignment policy, not an employment policy. This distinction is important when considering the extent to which Army women have been involved in direct combat or have been collocating with direct combat units because the policy does not constrain what women can do in the theater. The policy provides guidance about the specialties, positions, and units to which women can be formally assigned. It also provides guidance for Army leadership, force management, and assignment personnel to determine and code Army positions as gender-neutral or gender-specific and to distribute personnel to units. In doing so, the policy contains specific guidance for the following Army leadership personnel: the Secretary of the Army, the Assistant Secretary of the Army (Manpower and Reserve Affairs), the Deputy Chief of Staff for Personnel,[1] the Deputy Chief of Staff for Operations and Plans,[2] the chief of the National Guard Bureau, the chief of the Army Reserve, the commanding general of the Army's Training and Doctrine Command, and the commanders of the major commands.

That guidance focuses on the DCPC system to identify gender-limited billets and on replacement requisitioning and assignment practices. Individuals are assigned to the brigade level, at which the brigade

---

[1]   The Office of the Deputy Chief of Staff for Personnel is now referred to as G-1.

[2]   The Office of the Deputy Chief of Staff for Operations and Plans is now referred to as G-3.

commander and the brigade S-1 (personnel component) allocate them to specific jobs.[3] Thus, one might maintain that no one lower than the brigade commander in an operational unit is responsible for adhering to the assignment policy for individuals, as long as an entire unit (e.g., a company) is not thereafter assigned to a direct combat unit. This is established in the assignment policy as follows:

> Once properly assigned, female soldiers are subject to the same utilization policies as their male counterparts. In event of hostilities, female soldiers will remain with their assigned units and continue to perform their assigned duties.[4]

Thus, local commanders, especially tactical commanders, are free to use the resources they have in the most effective way they determine, as the policy states that female soldiers are subject to the same utilization policies as their male counterparts.

It is important to note that the assignment policy does not constrain what women can do individually once they arrive in a combat theater, but only to which positions or units they can be assigned. For example, some returnees reported that their units spent much of their time on METL tasks and the remaining time on general "soldier tasks," such as perimeter duty, guarding the gate, escorting locals, or cleaning the area. Other returning personnel stated that they spent only a portion of time on their METL tasks and that the rest of their time—in some instances, as much as 50 percent of their time—was spent on unanticipated tasks, and the extent and perceived danger of these additional duties varied considerably.

The extent to which a unit's activities are relevant to the assignment policy differs for the DoD and Army policies. The DoD direct combat restriction focuses on the primary mission of direct combat units. Thus, the doctrine of the unit, not its activities in the theater, will determine the units to which women can be assigned. The Army policy, how-

---

[3]   With the exception of command sergeant major, battalion commander, and brigade commander, who are selected centrally.

[4]   Headquarters, U.S. Department of the Army (1992, p. 2).

ever, includes restrictions (on collocation and direct combat) that require an assessment of the unit's activities. In the case of the direct combat restriction, the Army policy precludes women from being assigned to a unit whose routine mission includes direct combat. Because the routine activities of a unit might change without an accompanying change in doctrine, it is important to assess unit activities in the theater.

Opportunities Available to Army Women

Tables C.1, C.2, and C.3 list the occupations in the Army for enlisted personnel (by career management field [CMF]), warrant officers (by branch), and officers (by branch), respectively. Table C.1 also provides the pay grades of personnel in each of these occupations. All three tables note those occupations that are closed to women.

Table C.1
Enlisted MOS by CMF

MOS	Grade (low–high)	Title	Closed to Women
CMF 00, Immaterial			
00D	3-9	Special-duty area support group	
00F	3-9	MOS immaterial, National Guard Bureau	
00G	4-9	MOS immaterial, U.S. Army Reserve	
00S	4-9	Special-duty assignment, Armed Forces Staff College	
00Z	9-9	Command Sergeant Major	
14X	3-7	Space and missile defense operations	
CMF 09, Personnel Special Reporting Codes			
09B	1-7	Trainee, unassigned	
09C	1-5	Trainee, language	
09D	1-9	College trainee	

Table C.1—Continued

MOS	Grade (low–high)	Title	Closed to Women
09G	1-9	National Guard on active-duty medical hold	
09H	1-9	U.S. Army Reserve on active-duty medical hold	
09L	1-9	Interpreter/translator	
09N	1-7	Nurse corps candidate	
09R	1-9	Simultaneous membership	
09S	1-9	Commissioned officer candidate	
09T	1-9	College student, Army National Guard officer program	
09W	1-9	Warrant officer candidate	
11X	3-5	Infantry recruit	X
13X	1-4	Field artillery computer system specialist	X
18X	3-4	Special forces recruit	X
35W	3-4	Electronic warfare/signals intelligence recruit	
98X	3-4	Electronic warfare/signals intelligence recruit	
CMF 11, Infantry			
11B	3-7	Infantryman	X
11C	3-7	Indirect fire infantryman	X
11Z	8-9	Infantry senior sergeant	X
CMF 13, Field Artillery			
13B	3-7	Cannon crewmember	X
13C	3-7	Tactical automated fire control system specialist	X
13D	3-7	Field artillery automated tactical data system specialist	X
13E	3-6	Cannon fire direction specialist	X
13F	3-7	Fire support specialist	X

Table C.1—Continued

MOS	Grade (low–high)	Title	Closed to Women
13M	3-7	Multiple-launch rocket system/high-mobility artillery rocket system crewmember	X
13P	3-7	Multiple-launch rocket system operational fire direction specialist	X
13R	3-7	Field artillery firefinder radar operator	X
13S	3-7	Field artillery surveyor	
13W	3-7	Field artillery meteorological crewmember	
13Z	8-9	Field artillery senior sergeant	

CMF 14, Air Defense Artillery

MOS	Grade (low–high)	Title	Closed to Women
14E	3-7	Patriot fire control enhanced operator/maintainer	
14J	3-7	Air defense command, control, communications, computers, and intelligence tactical operations center enhanced operator/maintainer	
14M	3-7	Man-portable air defense system crew member (reserve component)	
14R	3-7	Bradley Linebacker crew member	X
14S	3-7	Air and missile defense crew member	
14T	3-7	Patriot launching station enhanced operator/maintainer	
14Z	8-9	Air defense artillery senior sergeant	

CMF 15, Aviation

MOS	Grade (low–high)	Title	Closed to Women
15B	3-6	Aircraft power plant repairer	
15D	3-6	Aircraft power train repairer	
15F	3-6	Aircraft electrician	
15G	3-6	Aircraft structural repairer	
15H	3-6	Aircraft pneudraulics repairer	
15J	3-7	OH-58D/ARH armament/electrical/avionics systems repairer	

Table C.1—Continued

MOS	Grade (low–high)	Title	Closed to Women
15K	7-7	Aircraft components repair supervisor	
15M	3-6	UH-1 helicopter repairer (reserve component)	
15N	3-6	Avionic mechanic	
15P	3-9	Aviation operations specialist	
15Q	3-7	Air traffic control operator	
15R	3-7	AH-64 attack helicopter repairer	
15S	3-7	OH-58D/ARH helicopter repairer	
15T	3-7	UH-60 helicopter repairer	
15U	3-7	CH-47 helicopter repairer	
15V	3-7	Observation/scout helicopter repairer (reserve component)	
15W	3-7	Unmanned aerial vehicle (UAV) operator	
15X	3-7	AH-64 armament/electrical/avionics systems repairer	
15Y	3-7	AH-64D armament/electrical/avionics systems repairer	
15Z	8-9	Aircraft maintenance senior sergeant	
CMF 18, Special Forces			
18B	5-7	Special forces weapon sergeant	X
18C	5-7	Special forces engineer sergeant	X
18D	5-7	Special forces medical sergeant	X
18E	5-7	Special forces communication sergeant	X
18F	7-7	Special forces assistant operations and intelligence sergeant	X
18Z	8-9	Special forces senior sergeant	X
CMF 19, Armor			
19D	3-7	Cavalry scout	X

Table C.1—Continued

MOS	Grade (low–high)	Title	Closed to Women
19K	3-7	M1 armor crewman	X
19Z	8-9	Armor senior sergeant	X
CMF 21, Engineer			
21B	3-7	Combat engineer	X
21C	3-7	Bridge crew member	
21D	3-8	Diver	
21E	3-5	Construction equipment operator	
21G	3-6	Quarrying specialist (reserve component)	
21H	6-7	Construction engineering supervisor	
21J	3-5	General construction equipment operator	
21K	3-5	Plumber	
21L	3-7	Lithographer	
21M	3-7	Firefighter	
21N	6-7	Construction equipment supervisor	
21P	4-7	Prime power production specialist	
21Q	3-7	Transmission and distribution specialist (reserve component)	
21R	3-5	Interior electrician	
21S	3-7	Topographic surveyor	
21T	3-7	Technical engineer	
21U	3-7	Topographic analyst	
21N	3-6	Concrete and asphalt equipment operator	
21W	3-5	Carpentry and masonry specialist	
21X	8-9	General engineering supervisor	
21Y	3-9	Terrain data specialist	
21Z	8-9	Combat engineering senior sergeant	

Table C.1—Continued

MOS	Grade (low–high)	Title	Closed to Women
CMF 25, Communication and Information System Operation			
25B	3-9	Information technology specialist	
25C	3-6	Radio operator/maintainer	
25D	3-7	Telecommunication operator-maintainer	
25F	3-6	Network switching system operator-maintainer	
25L	3-6	Cable system installer/maintainer	
25M	3-6	Multimedia illustrator	
25N	3-6	Nodal network system operator/maintainer	
25P	3-7	Microwave system operator/maintainer	
25Q	3-6	Multichannel transmission system operator/maintainer	
25R	3-6	Visual information equipment operator/maintainer	
25S	3-7	Satellite communication system operator/maintainer	
25T	8-8	Satellite/microwave system chief	
25U	3-8	Signal support system specialist	
25V	3-6	Combat documentation/production specialist	
25W	7-8	Telecommunication operation chief	
25X	9-9	Senior signal sergeant	
25Y	8-9	Information system chief	
25Z	7-9	Visual information operation chief	
CMF 27, Paralegal			
27D	3-9	Paralegal specialist	
CMF 31, Military Police			
31B	3-9	Military police	
31D	4-9	Criminal investigation division special agent	

Table C.1—Continued

MOS	Grade (low–high)	Title	Closed to Women
31E	3-9	Internment/resettlement specialist	

CMF 33, Military Intelligence System Maintenance/Integration

MOS	Grade (low–high)	Title	Closed to Women
33W	3-9	Military intelligence system maintainer/integrator	

CMF 35, Military Intelligence

MOS	Grade (low–high)	Title	Closed to Women
35F	3-8	Intelligence analyst	
35G	3-8	Imagery analyst	
35H	3-8	Common ground station analyst	
35K	3-8	UAV operator	
35L	5-8	Counterintelligence agent	
35M	3-8	HUMINT collector	
35N	3-7	Signals intelligence analyst	
35P	3-7	Cryptologic linguist	
35S	3-7	Signals collector/analyst	
35T	3-9	Military intelligence system maintainer/integrator	
35X	8-9	Intelligence senior sergeant/chief intelligence sergeant	
35Y	8-9	Chief counterintelligence/HUMINT sergeant	
35Z	8-9	Signals intelligence (electronic warfare) senior sergeant/chief	

CMF 36, Financial Management

MOS	Grade (low–high)	Title	Closed to Women
36B	3-9	Financial management technician	

CMF 37, Psychological Operations

MOS	Grade (low–high)	Title	Closed to Women
37F	3-9	PSYOP specialist	

CMF 38, Civil Affairs

MOS	Grade (low–high)	Title	Closed to Women
38B	3-9	Civil affairs specialist	

Table C.1—Continued

MOS	Grade (low–high)	Title	Closed to Women
CMF 42, Adjutant General			
42A	3-9		
42F	3-5		
42L	3-9		
42R	4-9		
42S	6-9		
CMF 44, Financial Management			
44C	3-9	Financial management technician	
CMF 46, Public Affairs			
46Q	3-6	Public affairs specialist	
46R	3-6	Public affairs broadcast specialist	
46Z	7-9	Chief public affairs noncommissioned officer (NCO)	
CMF 56, Religious Support			
56M	3-9	Chaplain assistant	
CMF 63, Mechanical Maintenance			
44B	3-5	Metal worker	
44E	3-7	Machinist	
45B	3-5	Small-arms/artillery repairer	
45G	3-5	Fire-control repairer	
45K	3-7	Armament repairer	
52C	3-6	Utility equipment repairer	
52D	3-6	Power-generation equipment repairer	
52X	7-7	Special purpose equipment repairer	
62B	3-7	Construction equipment repairer	
63A	3-7	M1 Abrams tank systems maintainer	X

Table C.1—Continued

MOS	Grade (low–high)	Title	Closed to Women
63B	3-6	Light-wheel vehicle mechanic	
63D	3-7	Artillery mechanic	X
63H	3-6	Track vehicle repairer	
63J	3-5	Quartermaster and chemical equipment repairer	
63M	3-7	Bradley fighting vehicle system maintainer	X
63X	7-7	Maintenance supervisor	
63Z	8-9	Mechanical maintenance supervisor	
CMF 68, Medical			
68A	4-9	Biomedical equipment specialist	
68D	3-7	Operating room specialist	
68E	3-8	Dental specialist	
68G	3-8	Patient administration specialist	
68H	3-7	Optical laboratory specialist	
68J	3-8	Medical logistics specialist	
68K	3-9	Medical laboratory specialist	
68M	3-8	Nutrition care specialist	
68P	3-8	Radiology specialist	
68Q	3-8	Pharmacy specialist	
68R	3-9	Veterinary food inspection specialist	
68S	3-9	Preventative medicine specialist	
68T	3-7	Animal care specialist	
68V	3-7	Respiratory specialist	
68W	3-8	Health care specialist	
68X	3-7	Mental health specialist	
68Z	9-9	Chief medical NCO	

Table C.1—Continued

MOS	Grade (low–high)	Title	Closed to Women
CMF 74, Chemical, Biological, Radiological and Nuclear (CBRN)			
74D	3-9	CBRN specialist	
CMF 79, Recruitment and Reenlistment			
79R	5-9	Recruiter	
79S	5-9	Career counselor	
79T	7-9	Recruiting and retention NCO (Army National Guard)	
79V	5-9	Retention and transition, NCO, U.S. Army Reserve	
CMF 88, Transportation			
88H	3-7	Cargo specialist	
88K	3-7	Watercraft operator	
88L	3-7	Watercraft engineer	
88M	3-7	Motor transport operator	
88N	3-7	Transportation management coordinator	
88P	3-7	Railway equipment repairer (reserve component)	
88T	3-7	Railway section repairer (reserve component)	
88U	3-7	Railway operations crew member (reserve component)	
88Z	8-9	Transportation senior sergeant	
CMF 89, Ammunition			
89A	3-5	Ammunition stock control and accounting specialist	
89B	3-9	Ammunition specialist	
89D	3-9	Explosive ordnance disposal specialist	
CMF 91, Mechanical Maintenance			
91A	3-7	M1 Abrams tank system maintainer	X

Table C.1—Continued

MOS	Grade (low–high)	Title	Closed to Women
91B	3-6	Light-wheel vehicle mechanic	
91C	3-6	Utility equipment repairer	
91D	3-6	Power-generation equipment repairer	
91E	3-6	Machinist	
91F	3-5	Small-arms/artillery repairer	
91G	3-5	Fire control repairer	
91H	3-6	Track vehicle repairer	
91J	3-5	Quartermaster and chemical equipment repairer	
91K	3-7	Armament repairer	
91L	3-6	Construction equipment repairer	
91M	3-7	Bradley fighting vehicle system maintainer	X
91P	3-7	Artillery mechanic	X
91W	3-5	Metal worker	
91X	7-7	Maintenance supervisor	
91Z	8-9	Mechanical maintenance supervisor	
CMF 92, Supply and Services			
92A	3-8	Automated logistical specialist	
92F	3-9	Petroleum supply specialist	
92G	3-9	Food service specialist	
92L	3-7	Petroleum laboratory specialist	
92M	3-9	Mortuary affairs specialist	
92R	3-9	Parachute rigger	
92S	3-9	Shower/laundry and clothing repair specialist	
92W	3-7	Water treatment specialist	
92Y	3-8	Unit supply specialist	

Table C.1—Continued

MOS	Grade (low–high)	Title	Closed to Women
92Z	9-9	Senior noncommissioned logistician	
CMF 94, Electronic Maintenance and Calibrations			
94A	3-7	Land combat electronic missile system repairer	
94D	3-7	Air traffic control equipment repairer	
94E	3-6	Radio and communication security repairer	
94F	3-6	Computer detection system repairer	
94H	3-7	Test, measurement, and diagnostic equipment maintenance support specialist	
94K	3-6	Apache attack helicopter system repairer	
94L	3-6	Avionic communication equipment repairer	
94M	3-6	Radar repairer	
94P	3-7	Multiple-launch rocket system repairer	
94R	3-6	Avionic system repairer	
94S	3-7	Patriot system repairer	
94T	3-7	Avenger system repairer	
94W	7-7	Electronic maintenance chief	
94X	7-7	Senior missile system maintainer	
94Y	3-7	Integrated family of test equipment operator/ maintainer	
94Z	8-9	Senior electronic maintenance chief	
CMF 96, Military Intelligence			
96B	3-8	Intelligence analyst	
96D	3-8	Imagery analyst	
96H	3-8	Common ground station operator	
96R	3-8	Ground surveillance system operator	X
96U	3-8	UAV operator	

Table C.1—Continued

MOS	Grade (low–high)	Title	Closed to Women
96Z	9-9	Intelligence senior sergeant	
97B	5-8	Counterintelligence agent	
97E	3-8	HUMINT collector	
97L	3-8	Translator/interpreter (reserve component)	
97Z	9-9	Counterintelligence/HUMINT senior sergeant	
CMF 98, Signals Intelligence/Electronic Warfare Operations			
98C	3-7	Signal intelligence analyst	
98G	3-7	Cryptologic communication interceptor/locator	
98P	3-7	Multisensor operator	
98Y	3-7	Signal collector/analyst	
98Z	8-9	Signal intelligence (electronic warfare), senior sergeant/chief	

Table C.2
Warrant Officer MOS by Branch

MOS	Title	Closed to Women
Branch 13, Field Artillery		
131A	Field artillery targeting technician	
Branch 14, Air Defense Artillery		
140A	Command-and-control system integrator	
140E	Patriot system technician	
140X	Air defense artillery, immaterial	
Branch 15, Aviation		
150A	Air Traffic and air space management technician	
150U	Tactical UAV operation technician	
151A	Aviation maintenance technician (nonrated)	

Table C.2—Continued

MOS	Title	Closed to Women
152B	OH-58A/C scout pilot (reserve component)	
152C	OH-6 pilot	
152D	OH-58D pilot	
152E	ARH-XX pilot	
152F	AH-64A attack pilot	
152G	AH-1 attack pilot (reserve component)	
152H	AH-64D attack pilot	
153A	Rotary-wing aviator (aircraft, nonspecific)	
153B	UH-1 pilot (reserve component)	
153D	UH-60 pilot	
153E	MH-60 pilot	
153L	LUH-XX pilot	
153M	UH-60M pilot	
154C	CH-47D pilot	
154E	MH-47 pilot	
154F	CH-47F pilot	
155A	Fixed-wing aviator (aircraft, nonspecific)	
155E	C-12 pilot	
155F	Jet aircraft pilot	
155G	0-5A/EO-5B/RC-7 pilot	
Branch 18, Special Forces		
180A	Special forces warrant officer	X
Branch 21, Corps of Engineers		
210A	Utility operation and maintenance technician	
215D	Geospatial information technician	

Table C.2—Continued

MOS	Title	Closed to Women
Branch 25, Signal Corps		
250N	Network management technician	
251A	Information system technician	
254A	Signal system support technician	
255Z	Senior signal system technician	
Branch 27, Judge Advocate General's corps		
270A	Legal administrator	
Branch 31, Military Police		
311A	Criminal investigation division special agent	
Branch 35, Military Intelligence		
350F	All-source intelligence agent	
350G	Imagery intelligence technician	
350K	UAV operation technician	
350Z	Attaché technician	
351L	Counterintelligence technician	
351M	HUMINT collection technician	
351Y	Area intelligence technician	
352N	Signal intelligence analysis technician	
352P	Voice intercept technician	
352Q	Communication interceptor/locator technician	
352R	Emanation analysis technician	
352S	Signal collection technician	
353T	Intelligence and electronic warfare system maintenance technician	
Branch 42, Adjutant General Corps		
420A	Human resource technician	

Table C.2—Continued

MOS	Title	Closed to Women
420C	Bandmaster	
Branch 64, Veterinary Corps		
640A	Veterinary service technician	
Branch 67, Medical Services Corps		
670A	Health service maintenance technician	
Branch 88, Transportation Corps		
880A	Marine deck officer	
881A	Marine engineering officer	
882A	Mobility officer	
Branch 89, Ammunition		
890A	Ammunition technician	
Branch 91, Ordnance		
913A	Armament system maintenance warrant officer	
914A	Allied trades warrant officer	
915A	Automotive maintenance warrant officer	
915E	Senior automotive maintenance officer/senior ordnance logistics officer	
919A	Engineer equipment maintenance warrant officer	
Branch 92, Quartermaster Corps		
920A	Property accounting technician	
920B	Supply system technician	
921A	Airdrop system technician	
922A	Food service technician	
923A	Petroleum technician	

Table C.2—Continued

MOS	Title	Closed to Women
Branch 94, Electronic Maintenance		
948B	Electronic system maintenance warrant officer	
948D	Electronics/missile maintenance warrant officer	
948E	Senior electronic system maintenance warrant officer	

Table C.3
Officer AOC by Branch

AOC	Title	Closed to Women
Branch 11, Infantry		
11A	Infantry	X
Branch 13, Field Artillery		
13A	Field artillery, general	
Branch 14, Air Defense Artillery		
14A	Air defense artillery, general	
14B	Short-range air defense artillery	
14D	Hawk missile air defense artillery	
14E	Patriot missile air defense artillery	
Branch 15, Aviation		
15A	Aviation, general	
15B	Aviation, combined armed operations	
15C	Aviation, all-source intelligence	
15D	Aviation, logistics	
Branch 18, Special Forces		
18A	Special forces	X
Branch 19, Armor		
19A	Armor, general	X

Table C.3—Continued

AOC	Title	Closed to Women
19B	Armor	X
19C	Cavalry	X
Branch 21, Corps of Engineers		
21A	Engineer, general	
21B	Combat engineer	
21D	Facility/contract construction management engineer	
Branch 25, Signal Corps		
25A	Signal, general	
Branch 27, Judge Advocate General Corps		
27A	Judge advocate general	
27B	Military judge	
Branch 31, Military Police		
31A	Military police	
Branch 35, Military Intelligence		
35C	Imagery intelligence	
35D	All-source intelligence	
35E	Counterintelligence	
35F	Human intelligence	
35G	Signals intelligence/electronic warfare	
Branch 37, Psychological Operations		
37A	PSYOP	
37X	PSYOP, designated	
Branch 38, Civil Affairs (Active Army and U.S. Army Reserve)		
38A	Civil affairs (active army and U.S. Army Reserve)	
38X	Civil affairs, designated	

Table C.3—Continued

AOC	Title	Closed to Women
Branch 42, Adjutant General Corps		
42B	Human resource officer	
42C	Army bands	
42H	Senior human resource officer	
Branch 44, Finance Corps		
44A	Finance, general	
Branch 56, Chaplain		
56A	Command and unit chaplain	
56D	Clinical pastoral educator	
Branch 60, Medical Corps		
60A	Operational medicine	
60B	Nuclear medicine officer	
60C	Preventative medicine officer	
60D	Occupational medicine officer	
60F	Pulmonary disease/critical care officer	
60G	Gastroenterologist	
60H	Cardiologist	
60J	Obstetrician and gynecologist	
60K	Urologist	
60L	Dermatologist	
60M	Allergist, clinical immunologist	
60N	Anesthesiologist	
60P	Pediatrician	
60Q	Pediatric subspecialist	
60R	Child neurologist	

Table C.3—Continued

AOC	Title	Closed to Women
60S	Ophthalmologist	
60T	Otolaryngologist	
60U	Child psychiatrist	
60V	Neurologist	
60W	Psychiatrist	
Branch 61, Medical Corps		
61A	Nephrologist	
61B	Medical oncologist/hematologist	
61C	Endocrinologist	
61D	Rheumatologist	
61E	Clinical pharmacologist	
61F	Internist	
61G	Infectious disease officer	
61H	Family medicine	
61J	General surgeon	
61K	Thoracic surgeon	
61L	Plastic surgeon	
61M	Orthopedic surgeon	
61N	Flight surgeon	
61P	Physiatrist	
61Q	Radiation oncologist	
61R	Diagnostic radiologist	
61U	Pathologist	
61W	Peripheral vascular surgeon	
61Z	Neurosurgeon	

Table C.3—Continued

AOC	Title	Closed to Women
Branch 62, Medical Corps		
62A	Emergency physician	
62B	Field surgeon	
Branch 63, Dental Corps		
63A	General dentist	
63B	Comprehensive dentist	
63D	Periodontist	
63E	Endodontist	
63F	Prosthodontist	
63H	Public health dentist	
63K	Pediatric dentist	
63M	Orthodontist	
63N	Oral and maxillofacial surgeon	
63P	Oral pathologist	
63R	Executive dentist	
Branch 64, Veterinary Corps		
64A	Field veterinary service	
64B	Veterinary preventative medicine	
64C	Veterinary laboratory animal medicine	
64D	Veterinary pathology	
64E	Veterinary comparative medicine	
64F	Veterinary clinical medicine	
64Z	Senior veterinarian (immaterial)	
Branch 65, Army Medical Specialist Corps		
65A	Occupational therapy	

Table C.3—Continued

AOC	Title	Closed to Women
65B	Physical therapy	
65C	Dietician	
65D	Physician assistant	
65X	Specialist allied operations	
Branch 66, Army Nurse Corps		
66B	Army public health nurse	
66C	Psychiatric/mental health nurse	
66E	Perioperative nurse	
66F	Nurse anesthetist	
66G	Obstetrics and gynecology	
66H	Medical-surgical nurse	
66N	Generalist nurse	
66P	Family nurse practitioner	
Branch 67, Medical Service Corps		
67A	Health service	
67B	Laboratory service	
67C	Preventative medicine service	
67D	Behavioral science	
67E	Pharmacy	
67F	Optometry	
67G	Podiatry	
67J	Aeromedical evacuation	
Branch 74, Chemical		
74A	Chemical, general	
74B	Chemical operations and training	

Table C.3—Continued

AOC	Title	Closed to Women
74C	Chemical munitions and materiel management	
Branch 88, Transportation Corps		
88A	Transportation, general	
88B	Traffic management	
88C	Marine and terminal operations	
88D	Motor/rail transportation	
Branch 89, Ammunition		
89E	Explosive ordnance disposal	
Branch 91, Ordnance		
91A	Maintenance and munitions materiel officer	
Branch 92, Quartermaster Corps		
92A	Quartermaster, general	
92D	Aerial delivery and materiel	
92F	Petroleum and water	

APPENDIX D
Army Women Deployed to Iraq

The RAND team requested and obtained from the Defense Manpower Data Center four data "snapshots" of Army personnel in Iraq: in April 2003, 2004, 2005, and 2006. The data included a unique identifier for each individual (a number) and his or her component, gender, pay grade, and MOS or area of concentration (AOC); the unit identification code (UIC) of the unit to which each individual was assigned; and the UIC of the unit in which each individual performed his or her duties (usually, these were the same). By matching the UICs with those in other Army databases, we were able to obtain the name and standard requirements code (SRC) of each unit. The SRCs, in turn, provide detailed information about the type of unit. For example, if the data showed that a female 92A (automated logistical specialist) was a sergeant in WJAXT0 we know that she was in the 64th Support Battalion, 3rd BCT, 4th Infantry Division, and that it was organized as a Force XXI FSB. The last two alphanumeric characters of the UIC indicate that this sergeant was in the battalion's headquarters company (HHC).

According to our data, the number of Army personnel in Iraq varied, from a high of 183,000 in April 2003, shortly after the conflict began, to a low of 135,000 the following April. In the past two years, the average number of personnel in Iraq at any given time has been approximately 150,000. Women comprised 11.5 percent of the total force in 2003 and around 10 percent in the following years. In other words, in each of the past three years, there have been close to 15,000 women in Iraq, and, in 2003, there were over 21,000.

The data presented in this appendix are from the April 2006 snapshot. It is appropriate to examine only a single snapshot—in this case, the most recent one—because the data show little evidence of any trends over time. The occupations and units in which women can be found have not changed since 2003, and the number of women in Iraq has remained relatively steady over the past three years. The unit data presented in this appendix are based on the soldiers' duty UIC, rather than on their assigned UIC.

Women in Pay Grades

In April 2006, women in Iraq filled all grades, from private (E-1) to brigadier general (O-7). Table D.1 lists the number and percentage of women in each grade. As the table shows, not all grades had equal proportions of women. This may be due to several factors, including the types of units in Iraq, the grade structure of the units in Iraq, and, for the warrant officer grades, the number of feeder MOSs from which warrant officers are drawn.

Table D.1
Women and Men in Iraq by Pay Grade, April 2006

Pay Grade	Women (n)	Men (n)	Percent Women
E-1	61	830	7
E-2	672	6,314	10
E-3	2,331	18,976	11
E-4	4,406	36,887	11
E-5	2,503	23,575	10
E-6	1,222	14,704	8
E-7	655	7,696	8
E-8	165	2,624	6

Table D.1—Continued

Pay Grade	Women (n)	Men (n)	Percent Women
E-9	51	811	6
Total enlisted	12,066	112,417	10
W-1	62	399	13
W-2	95	1,363	7
W-3	48	732	6
W-4	9	424	2
W-5	1	98	1
Total warrant officers	215	3,016	7
O-1	349	1,565	18
O-2	513	2,956	15
O-3	665	5,186	11
O-4	331	3,534	9
O-5	135	1,813	7
O-6	49	585	8
O-7	3	45	6
O-8	0	29	0
O-9	0	4	0
O-10	0	1	0
Total officers	2,045	15,718	12
Total personnel	14,326	131,151	10

Women in Occupations

The most common career fields among women in Iraq included logistics, medical occupations, military intelligence, and, for warrant officers, aviation. Table D.2 lists the CMFs, branches, and functional areas that comprised at least 3 percent of female enlisted soldiers, warrant officers, and officers.[1] Two out of every five enlisted women were in a logistics CMF. Supply and services, which comprised nearly a third of enlisted women, include individual MOSs, such as automated logistical specialist, petroleum supply specialist, unit supply specialist, and senior noncommissioned logistician. Similarly, a third of all female warrant officers were in the quartermaster corps. More interesting for the female warrant officers is the number of women in aviation MOSs. This CMF includes pilots of the Apache, Cobra, Kiowa, and Kiowa Warrior—all attack, close support, or reconnaissance helicopters. Women in these MOSs are certainly likely to engage enemy forces in combat, and these have been open to women for years. In April 2006, there were eight female Kiowa Warrior pilots and five female Apache pilots. A total of 31 women were rotary-wing pilots of

Table D.2
Most Common Occupations Among Women in Iraq, April 2006

CMF or Branch	Percent
Female enlisted	
Supply and services	31
Adjutant general	12
Medical	11
Transportation	8
Communication and information system operation	8
Mechanical maintenance	6
Military intelligence	4
Military police	3
CBRN	3
Total female enlisted (n)	12,252

---

[1]   We used the CMF, branch, and functional area groupings because there are more than 250 enlisted MOSs, nearly 100 warrant officer MOSs, and more than 200 officer AOCs. Very few individual MOSs or AOCs had a significant share of women.

Table D.2—Continued

CMF or Branch	Percent
Female warrant officers	
Quartermaster corps	33
Aviation	17
Military intelligence	10
Adjutant general's corps	9
Signal corps	8
Ordnance	6
Transportation corps	5
Military police	5
Total female warrant officers (n)	218
Female officers	
Army nurse corps	13
Quartermaster corps	10
Military intelligence	9
Signal corps	8
Ordnance	8
Medical service corps	7
Adjutant general corps	6
Transportation corps	6
Corps of engineers	6
Military police	4
Army medical specialist corps	3
Aviation	3
Medical corps	3
Total female officers (n)	2,097

some sort. Among female officers, the medical fields dominate, with more than a quarter of all female officers in one of the medical or health-related corps.

The preceding data and discussion examined the proportion of women in various CMFs and branches to answer the following question: If a woman is in Iraq, what is she likely to be doing? The same data can be used to examine the proportion of female personnel in each CMF and branch to answer the following question: How likely is it that a particular job is being done by a woman? In answer to this second question, the CMFs with the largest proportion of enlisted women included administration, the adjutant general corps, financial management, public affairs, and the medical corps. These CMFs were usually at least 20 percent female, with a few exceeding 30 percent. The warrant officer branches with the highest proportion of women included the adjutant general, quartermaster, transportation, signal, and medical corps. The proportion of women in these warrant officer branches did not exceed 30 percent. Among officers, the occupations with the highest proportion of women included many of the health-related branches, finance, the adjutant general corps, and logistics. Only in 2005 did the Army nurse corps dip below 50 percent female—it has the highest proportion of women by far. But several of the other branches and functional areas were 30 to 40 percent female. Of course, across all ranks, the combat arms branches had very few women. Aviation tended to have the most, but that proportion was only about 5 or 6 percent. Tables D.3 through D.5 give the percentage of women in the occupational groupings for enlisted, warrant officer, and commissioned officers, respectively.

Table D.3
Women and Men in Iraq by Enlisted CMF, April 2006

CMF	Women (n)	Men (n)	Percent Women
Administration[a]	41	77	35
Adjutant general	1,439	2,787	34
Financial management	187	380	33
Paralegal	106	246	30
Public affairs	71	196	27
Supply and services	3,699	12,383	23
Bands[a]	2	7	22
Medical	1,328	5,365	20
Religious support	57	248	19
Ammunition	193	928	17
Recruitment and reenlistment	33	161	17
CBRN	312	1,659	16
Signals intelligence/electronic warfare operations[a]	119	663	15
Civil affairs	62	355	15
Transportation	1,020	5,946	15
Military intelligence	454	2,715	14
Communication and information system operation	932	7,434	11
Military police	417	3,428	11
PSYOP	50	423	11
Petroleum and water[a]	7	61	10
Electronic maintenance and calibration	121	1,073	10
Personnel special reporting codes	11	166	6
General engineering[a]	4	61	6
Aviation	284	4,382	6

Table D.3—Continued

CMF	Women (n)	Men (n)	Percent Women
Mechanical maintenance	783	13,254	6
CMF immaterial	21	398	5
Engineer	266	7,429	3
Military intelligence system maintenance/ integration[a]	6	202	3
Air defense artillery	10	811	1
Field artillery	12	10,746	0
Armor	1	6,369	0
Aircraft maintenance[a]	0	29	0
Aviation operations[a]	0	3	0
Combat engineering[a]	0	147	0
Infantry	0	20,086	0
Special forces	0	1,441	0

[a] Scheduled to be deleted, but still appears in the personnel database.

Table D.4
Women and Men in Iraq by Warrant Officer CMF, April 2006

CMF	Women (n)	Men (n)	Percent Women
Adjutant general's corps	19	58	25
Quartermaster corps	70	219	24
Transportation corps	11	53	17
Veterinary corps	1	5	17
Signal corps	17	99	15
Military police	10	64	14
Ammunition	4	28	13
Military intelligence	22	181	11

Table D.4—Continued

CMF	Women (n)	Men (n)	Percent Women
Judge advocate general's corps	1	11	8
Electronic maintenance	4	48	8
Medical service corps	1	15	6
Ordnance	13	441	3
Aviation	36	1,458	2
Field artillery	0	114	0
Air defense artillery	0	20	0
Special forces	0	97	0
Corps of engineers	0	27	0

Table D.5
Women and Men in Iraq by Officer Branch and Functional Area, April 2006

Branch or Functional Area	Women (n)	Men (n)	Percent Women
Army nurse corps	274	212	56
Veterinary corps	19	24	44
Adjutant general corps	129	212	38
Behavioral sciences	7	16	30
Finance corps	38	103	27
Medical service corps	140	469	23
Logistics	4	15	21
Army medical specialist corps	60	231	21
Health services	28	109	20
Space operations	2	9	18
Quartermaster corps	198	906	18
Signal corps	172	859	17
Chemical	49	248	16

Table D.5—Continued

Branch or Functional Area	Women (n)	Men (n)	Percent Women
Military police	82	426	16
Judge advocate general's corps	48	250	16
Military intelligence	180	938	16
Ordnance	155	818	16
Preventive medicine sciences	3	16	16
Human resource management[a]	4	23	15
Transportation corps	121	723	14
Medical corps	56	409	12
Dental corps	7	62	10
Foreign area officer	5	45	10
System engineering	1	9	10
Corps of engineers	120	1,197	9
Strategic intelligence	1	10	9
Civil affairs (active and reserve)	18	190	9
Information operations	2	23	8
Research, development, and acquisition	11	127	8
Aviation	60	864	6
Public affairs	2	35	5
Air defense artillery	15	273	5
Strategic plans and policy	1	19	5
Operations research/system analysis	1	22	4
Chaplain	14	321	4
Comptroller	1	29	3
PSYOP and civil affairs[a]	2	62	3
Armor	1	1,077	0

Table D.5—Continued

Branch or Functional Area	Women (n)	Men (n)	Percent Women
Field artillery	1	1,627	0
Ammunition	0	9	0
Force development	0	6	0
Infantry	0	2,050	0
Laboratory sciences	0	3	0
Nuclear and counterproliferation	0	3	0
Simulation operations	0	10	0
Special forces	0	328	0
Systems automation officer	0	30	0
U.S. Military Academy stabilized faculty	0	2	0

a Scheduled to be deleted, but still appears in the personnel database.

Women in Units

The Army's basic tactical maneuver unit is the BCT. According to Army doctrine, the core mission of the BCTs is "to close with the enemy by means of fire and maneuver to destroy or capture enemy forces, or to repel their attacks by fire, close combat, and counterattack."[2] That description is close to the definition of *direct ground combat* that shapes the Army's assignment policy, though the assignment policy pertains to units that are battalion size or smaller. This is significant because more than 2,000 of the 14,000 women in Iraq as of April 2006 were performing their duties as part of a BCT.

Today's HBCTs and infantry brigade combat teams (IBCTs) typically have three maneuver battalions or squadrons, a field artillery battalion, a support battalion, a special troops battalion, and an HHC. The HBCTs are typically authorized about 3,800 troops and IBCTs

---

2 Headquarters, U.S. Department of the Army (2006b, p. 2-1).

are typically authorized about 3,450. The average BCT at or near full strength in Iraq in April 2006 had about 170 women.

The field artillery and maneuver battalions within BCTs are closed to women. About three-quarters of the positions in the HHCs are open to women, and the ones that are closed are such due to the branch or MOS, not the mission of the subunit. These are brigade staff jobs, and the HHCs are small, with only about 60 or so total personnel. The majority of the women in a BCT are found in the support battalion and the special troops battalion. Obviously, the Army's interpretation of its assignment policy does not preclude women from being assigned to BCTs themselves; rather, the assignment policy is applied to lower-echelon units within the BCTs, namely the field artillery and maneuver battalions and squadrons.

Brigade Support Battalions

The typical BSB is organized as shown in Table D.6. Given the mission of the FSCs, they are clearly in close proximity to the maneuver units whose mission is to engage the enemy in direct ground combat.

On paper, the typical HBCT BSB is organized with about 1,200 troops, and the typical IBCT BSB is organized with about 870 troops.[3] In actuality, no BSB had more than 1,050 on the ground in Iraq in April 2006, and, more typically, there were between 500 and 700 troops in a single BSB. Between 40 and 50 of the 1,200 authorizations are closed to women, and those are in just three restricted MOSs. Most of the positions in each company, and even in each platoon, are open to women, though there are some squads that are comprised entirely of restricted MOSs and are thus closed to women.

In practice, about 15 to 25 percent of BSB troops are female. The duty UIC data show women in every type of company within a BSB, including the FSCs that support maneuver units.[4] For example, the

---

[3] Data provided by U.S. Army Force Management Support.

[4] The company-level information (typically indicated by the last two characters of the six-character UIC) is less reliable than the battalion-level information (the first four characters), but other information corroborates the data on women in the FSCs.

Table D.6
Organization of BSBs in HBCTs and IBCTs

Subunit	Mission
HHC	Provide command and control for all organic and attached units of the BSB.
Distribution company	Provide transportation and supply support to the HBCT.
Field maintenance company	Provide field-level maintenance support to a BSB.
Medical company	Provide echelon II medical care to supported maneuver battalions with organic medical platoons. Provide echelon I and II medical treatment on an area basis to units without organic medical assets operating in the brigade area of operations (AO).
FSC (armed reconnaissance squadron [ARS])	Provide direct and habitual combat service support to itself and the ARS.
2 FSCs (maneuver battalion)	Provide direct and habitual combat service support to itself and the maneuver battalion.
FSC (field artillery battalion)	Provide direct and habitual combat service support to itself and the field artillery battalion.

SOURCE: Headquarters, U.S. Department of the Army, Brigade Support Battalion (HBCT) Table of Organization and Equipment Narrative, DOCNO 63325GFC05, effective May 16, 2006a.

64th Support Battalion, 3rd BCT, 4th Infantry Division, had 1,050 troops in Iraq in April 2006. This was the largest of the BSBs in terms of number of people in Iraq. As Table D.7 shows, the company-level UIC data indicate that there were enlisted women in each of the battalion's eight companies. Six of the eight companies had female officers, and two of the eight had female warrant officers. Women comprised 15 percent of all enlisted troops, 14 percent of warrant officers, and 23 percent of commissioned officers. However, the proportion of women in the FSCs was smaller—less than 10 percent. This is representative of the demographics of the other BSBs.

Table D.7
Composition of the 64th Support Battalion, 3rd Brigade Combat Team,
4th Infantry Division in Iraq, April 2004

Company	Enlisted (n)		Warrant (n)		Officer (n)		Total
	F	M	F	M	F	M	
Headquarters	18	32	0	3	5	15	73
Distribution	27	96	1	1	0	7	132
Field maintenance	19	96	1	3	1	4	124
Medical	26	32	0	0	6	7	71
Forward support (ARS)	11	113	0	1	1	5	131
Forward support (maneuver battalion)	16	174	0	1	0	3	194
Forward support (maneuver battalion)	16	175	0	1	1	3	196
Forward support (field artillery)	6	94	0	1	1	4	106
Unknown	6	15	0	1	0	1	23
Total	145	827	2	12	15	49	1,050

The Stryker brigade combat Teams (SBCTs) are also organized around three maneuver battalions, with one BSB for support. They do not, however, have a special troops battalion. Although the SBCTs are slightly larger than the HBCTs, their support battalions are only about half the size of those of the HBCTs. But in terms of the percentage of the unit that is female and the distribution of women across the companies, the support battalions of the SBCTs are about the same as those of the other BCTs.

Fire and aviation brigades are organized somewhat differently, but they, too, have BSBs, typically with several hundred troops. Those support battalions are also in the neighborhood of 20 percent female, like the BSBs in the BCTs. For example, the 96th Support

Battalion, 101st Aviation Brigade, 101st Airborne Division, was one of the largest aviation support battalions in Iraq in April 2006. All four companies had female enlisted troops, one had a female warrant officer, and three of the four had female officers. Women comprised 15 percent of all enlisted troops, 11 percent of warrant officers, and 32 percent of commissioned officers.

Overall, it would appear that about one in five people in any given BSB is female, and they can be found throughout the various units and subunits of the BSBs, including FSCs that provide habitual support to the ground combat battalions.

Women in Brigade Special Troops Battalions

A relatively new type of unit found in the HBCTs and IBCTs is the brigade special troops battalion (BSTB), which is organized as shown in Table D.8. The combat engineer company is closed to women because it sometimes performs infantry combat missions, and it is found only in the IBCTs. The BSTBs of the HBCTs do not have engineer companies.

Although the BSBs tended to be smaller on the ground than their authorized strength, the BSTBs were nearly at full strength. An IBCT BSTB has an authorized strength of 398, and an HBCT BSTB has an authorized strength of 314. Many of the BSTBs in Iraq in April 2006 were at or near these levels. Overall, women comprised about 10 to 15 percent of the BSTBs in the IBCTs and 15 to 20 percent in the HBCTs. The IBCT BSTBs had a lower percentage of women because of the all-male engineer companies.

Table D.9 shows the composition of the BSTB of the 4th BCT, 101st Airborne Division, in April 2006. As noted, there were no women in the engineer company, but women—enlisted as well as officers—were found in each of the other companies. There was also one female warrant officer in the HHC. These numbers are representative of the other BSTBs.

Table D.8
Organization of BSTBs in HBCTs and IBCTs

Subunit	Mission
HHC	Provide command, control, and supervision of battalion tactical operations. Provide unit administration and logistical support for the battalion staff sections. Provide administrative, logistical, and medical support to organic and attached units.
Engineer company (IBCTs only)	Increase the combat effectiveness of the separate infantry brigade by accomplishing limited mobility, countermobility, survivability, and sustainment engineering missions. Perform infantry combat missions when required.
Military intelligence company	Provide timely, relevant, accurate, and synchronized intelligence, surveillance, and reconnaissance support to maneuver units within the BCT and BCT commander, staff, and subordinates during the planning, preparation, and execution of multiple, simultaneous decision actions on a distributed battlefield.
Signal network support company	Deploy, install, operate, and maintain the brigade's command, control, communications, computer, intelligence, surveillance, and reconnaissance network. Establish networks that support brigade operations and integrate with division army force, joint task force, or theater networks.

SOURCE: Headquarters, U.S. Department of the Army, Brigade Special Troops Battalion (HBCT) Table of Organization and Equipment, Narrative, DOCNO 87305GFC12, effective October 16, 2005b.

Conclusions

During our review of the roles and missions of many different types of Army units, we found several units that (1) would seem to require collocation or habitual association with ground combat units to fulfill their missions and (2) had a significant number of positions open to women. That naturally invites the question, Are those positions open to women but in practice filled by men, or do women actually fill the positions? The brief analysis provided here has answered that question: Women are found in almost any unit or subunit that is open to them

Table D.9
Composition of the BSTB, 4th BCT, 101st Airborne Division in Iraq,
April 2006

Company	Enlisted (n)		Warrant (n)		Officer (n)		Total
	F	M	F	M	F	M	
Headquarters	27	117	1	0	2	15	162
Engineer	0	71	0	0	0	5	76
Military intelligence	11	65	0	2	2	2	82
Signal network support	8	47	0	1	1	2	59
Total	46	300	1	3	5	24	379

within a BCT, and they are not there in just ones and twos. The specific units whose male-female composition ratio was listed in the tables in this appendix were selected because they are representative of typical units in Iraq.

Interviews with Senior Army, OSD, and JS Personnel and Members of Congress

This research included 11 semistructured interviews with senior Army, OSD, and JS personnel, as well as less structured interviews and discussions with congressional staffers and members of Congress. The purpose of the DoD interviews was threefold: to determine the extent to which DoD leadership understood and shared a common understanding of the precise meaning of the assignment policy, to determine what DoD leadership felt were the objectives of the current assignment policy, and to determine what they felt were appropriate objectives for an assignment policy. The interviews on Capitol Hill focused almost exclusively on the latter two of these; we did not spend time discussing the wording or the implementation of the current policy. The following is a list of senior DoD personnel interviewed:

- David S. C. Chu, Under Secretary of Defense for Personnel and Readiness
- Paul W. Mayberry, Deputy Under Secretary of Defense for Readiness
- BG K. C. McClain, Commander, Joint Task Force for Sexual Assault Prevention and Response, Office of the Under Secretary of Defense for Personnel and Readiness
- RADM Donna Crisp, Joint Staff, Personnel (J1)
- LTG Walter L. Sharp, Director, Joint Staff
- LTG James L. Campbell, Director, Army Staff

- Kathryn A. Condon, Assistant Deputy Chief of Staff, Office of the Deputy Chief of Staff for Operations (G-3/5/7)
- MG Rhett Hernandez, Commander, U.S. Army Human Resources Command
- Mark R. Lewis, Deputy Chief of Staff, Office of the Deputy Chief of Staff for Personnel (G-1)
- Mark D. Manning, Special Assistant to the Assistant Secretary of the Army (Manpower and Reserve Affairs)
- BG Robert H. Woods, Director, Enlisted Personnel Management Directorate

Objectives Considered in the Interviews

Each of the objectives considered in the interviews is discussed briefly below. These objectives emerged during the ongoing discussions of women in the military:

- *Maximize the effectiveness of the military.* This objective reflected the inquiry regarding whether constraints on women's roles in the military supercede military effectiveness.
- *Maximize the flexibility of assigning people.* This objective was intended to reflect whether efficiency and flexibility of military processes could coincide with prescribed roles for women.
- *Maintain current opportunities for women.* This objective addressed whether the current opportunities for military women are acceptable.
- *Open new career opportunities for women.* This objective reflected attitudes that current opportunities for women are insufficient.
- *Provide career opportunities to make women competitive with their male counterparts in career advancement.* This objective reflected concerns that women might not be competitive with their male peers in career advancement if they were not to have the opportunities to obtain important job assignments.

- *Protect female service members from physical harm.* This objective addressed the belief held by some that the assignment policy was intended to shield female service members from harm.
- *Protect male service members from physical harm by excluding women from ground combat.* This objective reflected attitudes that female presence among ground combat personnel or in units might lessen the combat arms' effectiveness or otherwise endanger male combat arms personnel.
- *Simplify unit leadership by limiting male-female interaction.* This objective reflected the assertion that commanding combat arms units is sufficiently difficult without also introducing gender issues.
- *Exclude women from ground combat occupations and units.* This objective reiterates that the purpose of an assignment policy is to specifically preclude women from ground combat occupations and units.
- *Exclude women from occupations that require considerable physical strength.* This objective reiterates concerns that women should not be assigned to occupations that require considerable physical strength, regardless of the proximity of those assignments to combat.
- *Ensure buy-in from all involved parties and stakeholders through compromise.* This objective reflects the opinion that the current assignment policy was established through a difficult process of compromise and that any future policy should also have the agreement of such parties as Congress, OSD, and the services.

Protocol Used in Interviews with DoD Leadership

This section presents the protocol followed in the interviews with senior Army, OSD, and JS personnel who participated in the interviews conducted for this study.

You were selected to participate in our study to provide the Army/OSD viewpoint regarding the assignment policy for women. Accordingly, as we go through our questions today, please answer them from the perspective of the Army/OSD. Our interview consists of open-ended questions, as well as some structured questions.

1. What are the objectives of the existing assignment policy for women?

Prompts:
- What is the intent of the policy?
- Why was this assignment policy developed?
- Why was this policy instituted, rather than just eliminating the risk-based policy and opening everything to women?
- Why do positions and units remain closed to women?

2. How well does the current policy meet the existing objectives?

3. Are these the right objectives for an ideal policy, or would you add or delete objectives?

4. Part of the first task of our research is to understand institutional perspectives on various objectives of the assignment policy.

[Interviewee is provided with a page that lists items 4a through 4k.]

For each of the 11 potential assignment policy objectives on the sheet, please let us know which of the selections from the five-point scale at the top of the sheet you would use to indicate the extent to which OSD/JS/Army would agree or disagree that it is important for an assignment policy to address that particular objective (you need not be constrained by the current policy). You do not need to write anything; I will record your answers.

4a. It is important that an assignment policy for women maximizes the operational effectiveness of the military.
4b. It is important that an assignment policy for women maximizes the flexibility of assigning people.

4c. It is important that an assignment policy for women maintains current opportunities for women.

4d. It is important that an assignment policy for women provides career opportunities to make women competitive with their male counterparts in career advancement.

4e. It is important that an assignment policy for women opens new career opportunities for women.

4f. It is important that an assignment policy for women protects female service members from physical harm.

4g. It is important that an assignment policy for women protects male service members from physical harm by excluding women from ground combat.

4h. It is important that an assignment policy for women simplifies unit leadership by limiting male-female interaction.

4i. It is important that an assignment policy for women excludes women from ground combat occupations and units.

4j. It is important that an assignment policy for women excludes women from occupations that require considerable physical strength.

4k. It is important that an assignment policy for women be an act of compromise among all parties and stakeholders involved. (If yes, ask, "Who are the parties involved?")

5. The next, related series of questions addresses the extent to which the current assignment policy for women successfully addresses those objectives.

[Interviewee is provided with a page that lists items 5a through 5i.]

The objectives are listed on the second sheet. Please use the same scale as before to indicate Army/OSD/JS agreement or disagreement with the extent to which the current assignment policy satisfies that objective.

5a. The current assignment policy for women successfully maximizes the operational effectiveness of the military.

5b. The current assignment policy for women successfully maximizes the flexibility of assigning people.

5c. The current assignment policy for women successfully provides career opportunities to make women competitive with their male counterparts in career advancement.

5d. The current assignment policy for women successfully protects female service members from physical harm.

5e. The current assignment policy for women successfully protects male service members from physical harm by excluding women from ground combat.

5f. The current assignment policy for women successfully simplifies unit leadership by limiting male-female interaction.

5g. The current assignment policy for women successfully excludes women from ground combat occupations and units

5h. The current assignment policy for women successfully excludes women from occupations that require considerable physical strength

5i. The current assignment policy for women successfully reflects an act of compromise among all parties and stakeholders involved. (If yes, ask, "Who are the parties involved?" [unless asked for 4k].)

I'm going to move back to a few open-ended questions now. Those were all the structured questions I have today.

6. From an Army/OSD/JS perspective, what are the advantages and disadvantages of the current assignment policy?

My last few questions pertain to the specifics of the current policy:

7. How close do two units have to be to one another to be considered physically collocated?
Prompt if a vague answer is given:
  • How do you know if units are collocated?

8. The current policy mentions *physically collocate and remain*. How long do you have to be with a direct ground combat unit before you have "remained"?

Probe:
- What about frequency? Does it matter how often you spend time with them? (For example, if you spend only a few hours or daylight hours with them, but you do so everyday, is that "remaining"?)

9. What does it mean to be *well forward* in Iraq?

Prompt, if interviewee cannot define well forward:
- The definition of *direct ground combat* says, "Direct ground combat is engaging an enemy on the ground with individual or crew served weapons, while being exposed to hostile fire and to a high degree of direct physical contact with the hostile force's personnel. Direct ground combat takes place well forward on the battlefield while locating and closing with the enemy to defeat them by fire, maneuver, or shock effect."

Probe, as necessary:
- Absent a definition of *well forward*, what is the definition of *direct ground combat*?

10. The assignment policy uses the term *enemy*. How difficult is it to define *enemy* in an area such as Iraq?

Probe, as necessary:
- Why? How so?

11. How does the Army/OSD/JS define when service members are *closing with the enemy* in the context of a counterinsurgency?

Probe, as necessary:
- Said another way, how close do service members have to be before they are *closing with the enemy*?
- Are there factors other than distance that affect whether one is *closing with the enemy*?

12. Since the Army definition of direct combat includes both offensive and defensive engagements, how does the Army/

OSD/JS define *repelling the enemy's assault* in the context of a counterinsurgency?

13. How do commanders know [or determine] whether they are exposing personnel to hostile fire?

14. Is there anything that you would like to add before we conclude?

Interviews and Focus Groups with Personnel Recently Returned from Iraq

The research team conducted interviews and focus groups with personnel who had recently returned from Iraq during two visits to Army locations. The first trip was to Fort Lee, Virginia, where the research team conducted four focus groups, three with groups of Combined Logistics Captains Career Course (CLC3) students and one with selected CLC3 instructors. In total, 19 officers participated in these focus groups. These individuals had deployed to Iraq as company-grade officers in a variety of occupations, including combat arms, combat support, and CSS. Most, but not all, of the participants were captains.

The second fieldwork visit was to a recently returned Army unit. This visit provided the majority of the focus groups. We conducted focus groups with the more junior personnel and interviews with battalion and brigade command.

In total, 80 people from these two locations participated in 16 focus groups and eight interviews. At least one interview or focus group was conducted with each of the following types of personnel:

- FCS company commanders and first sergeants
- platoon leaders from maintenance, supply, transportation, and medical units
- BSTB commanders
- BCT commander, executive officer, and command sergeant major
- maneuver battalion commander and executive officer

- BSB/aviation support battalion commanders
- female sustainment brigade personnel
- platoon leaders and platoon sergeants from military police units
- female FSC personnel
- female personnel from military police units
- female personnel from military intelligence units
- female company-grade officers.

Table F.1 presents characteristics of the personnel who participated in the focus groups and interviews.

Recruiting Participants and Conducting the Focus Groups and Interviews

For both visits, the Army identified a point of contact who organized the visit, scheduled the focus groups and interviews, and coordinated participant selection. In the case of the unit visit, the Army also selected the location, based on the availability of recently returned units. RAND did not select individual participants, although we did provide preferred participant characteristics for each group. Our request for different kinds of groups was designed to gather individuals with similar unit experience while keeping pay grade groups largely

Table F.1
Focus Group and Interview Participants

Participant Characteristic	Number in Focus Groups	Number in Interviews
Male	37	7
Female	34	2
Total	71	9
Enlisted	30	2
Officer	41	7
Total	71	9

separate. We requested some all-female groups to ensure that we captured the experiences of women, but other groups were requested based on specific jobs (e.g., company commanders and first sergeants). Some of the groups requested by job were all-male groups. There were instances in which we requested a group with an insufficient number of personnel. For example, we requested a group of female NCOs from FSCs. The focus group was instead populated with both NCOs and junior enlisted women, due to the very small number of female NCOs. Thus, the list of personnel types presented earlier in this appendix reflects the interviews and focus groups conducted, but it is roughly similar to the list of sessions requested. We also specified the types of units from which we wanted to interview the commander. As a result, we interviewed the commander and/or executive officer of a brigade combat team, a maneuver battalion, a BSB, and a BSTB.

We averaged about four participants in each focus group; the largest group had seven. On occasion, a participant would arrive late or a group would include fewer participants than we had expected and we would conduct these sessions as interviews, using the same protocol as for the focus groups. Participants were assured of confidentiality and were not compensated for their time.

The interviews and focus groups lasted approximately 60 minutes and were conducted by two experienced RAND researchers, one of whom led the session while the other took notes. The notes were subsequently transcribed, entered into a database using qualitative coding software, and analyzed with grounded theory.

Protocol for Interviews with Battalion and Brigade Leadership

The questions presented here were adjusted slightly to accommodate the different types of units. Nonetheless, the following is the basic protocol used for interviews with commanders. The protocol used for focus groups with returned personnel follows.

1. Can you confirm the kind of unit that you deployed with to Iraq and your role in that unit?

2. Can you tell me about your unit's primary mission or missions in Iraq and, roughly, where you were?

3. While I'm sure this will differ across your unit, can you estimate the amount of time that your personnel generally spent on METL tasks and on non-METL tasks?

4. How appropriate was doctrine to what units were actually doing in Iraq? Asked another way, is doctrine being revised to reflect what worked in Iraq?

5. How close did the FSC or other support personnel tend to be to the maneuver units? [Request approximate distance.] [Ask of BTB commanders:] How close did your personnel tend to be to the maneuver units? [Request approximate distance.]

6. Did support personnel travel outside the FOB to maneuver units, and, if so, how long did they tend to stay?

Probes (if necessary):
- What kind of support personnel?
- Why were those longer stays necessary?
- Were both male and female personnel involved in those longer stays?
- Would you characterize any of those interactions as high risk?

7. What kind of relationship did maneuver units have with support units? For example, were any of the support units attached to maneuver units?

8. Did you make or see any decisions made to treat women differently from their male peers?

Probes:
- For example, were men purposely chosen for tasks over their female peers?

- Were there any efforts to reduce the risk to women specifically?
- [If necessary] Such as what?
- Who made these decisions?

9. Did you need to use support personnel in any way that seemed to include tasks outside those traditionally associated with their MOS?

Probes (if needed):

- For example, were any signal or intelligence personnel going door to door with combat arms personnel?
- Were female support personnel used to search local female civilians?

10. Are you familiar with the assignment rule for military women? [If no, read the Army version.] Does the rule make sense to you, given operations in Iraq?
    [If yes, ask, "How do you define terms like *far forward, collocate and remain*, and *enemy*?"

Protocol for Focus Groups

[The notetaker records pay grade, sex, and any other background information provided.]

   [All participants respond to the first question in turn.]

1. When were you deployed to Iraq, what kind of a unit were you in, and what was your job in that unit?

[Unit size will depend on the individual, but comments sought pertain to the unit at the participant's pay-grade level.]

2. How many of you were in units that supported a BCT?

[Notetaker records the number of hands raised.]

3. Roughly how many or what proportion of the personnel in your unit were female, and in which occupations and pay grades? I don't need a lot of detail, but it would help me to know where the women were located in your unit and what they were doing?

[An answer is recorded for each participant; order is not important. The total personnel in the unit is noted.]

4. What were your unit's primary missions and what METL tasks did you conduct to support these missions?

5. How much time was spent on your METL tasks and how much time was spent on other things?

6. What were those non-METL tasks? Did they involve high risk or danger?

7. How close or far away did you tend to be from the combat arms units? [Request approximate distance.]

8. What kind of relationship did your unit have with a combat unit? For example, was it attached or supporting? If supporting, was it habitual?

9. Please describe the interaction your unit had with the combat arms units. For example, how often did your unit interact with the combat arms units and for what purpose? How often did you travel to their locations? How long did you tend to stay, and did you have longer stays with them?

Probes (if necessary):
- Why were those longer stays necessary?
- Were both male and female personnel involved in those longer stays?

10. Did combat arms personnel provide security for your unit? When did they provide security?

Probes (if necessary):
- For example, when in an FOB, in convoys, on recovery missions, etc.?
- Did you have a habitual security element assigned?

11. Were there any decisions made to treat women in your unit differently from their male peers?

Probes (if necessary):
- For example, did your unit ever keep women in a safer zone while male personnel went out, or were there concerns about women operating in dangerous areas or around combat arms units?
- Were women involved in any of the non-METL tasks you mentioned earlier?
- Did the unit take actions to reduce the risk to women? [If yes, ask, "What actions? Why?"]
- Whose decision was that?

12. Were there any performance differences between the men and the women in your unit?

13. Are you familiar with the assignment rule for military women? [If no, read the Army assignment policy.] Does the rule make sense to you, given operations in Iraq?
[If yes, ask, "How do you define terms like *far forward, collocate and remain,* and *enemy*?"]

14. Were any of your METL tasks carried out in unconventional ways or in ways inconsistent with Army doctrine?
[If yes, ask, "Do you think that any of those changes contributed to greater risk or helped you to minimize risk?"]

Army Modularity, Asymmetric Threats, and Nonlinear Battlefields

This appendix is intended to provide additional explanation of the Army's transformation to modular units, asymmetric military threats, and the difference between linear and nonlinear battlefields.

The Army's Modularity Conversion

To divest cold-war structure and improve the deployability and readiness of units, the Army has been undergoing its most major reshaping in decades. As one part of this reshaping, the Army has moved from a division-based concept of organization to a brigade-centric focus. Previously, when the Army deployed forces, it would deploy large divisions. These 10 divisions contained numerous maneuver brigades, artillery brigades, and various combat support and combat service support brigades. Needless to say, divisions were very large and could take a significant amount of time and strategic assets to deploy. In addition, not all deployments required all the elements or the number of forces in a division (e.g., Bosnia deployments). Therefore, the Army's brigade-centric restructuring provides DoD a larger number of deployable units (e.g., approximately 40 BCTs, as opposed to 10 divisions, are in the pool of forces), which are more agile and self-contained and can be better tailored to future missions, as opposed to a cold-war enemy. These goals were accomplished, in part, by moving, eliminating, or restructuring elements and capabilities that previously were contained in (i.e., organic to) divisions into BCTs directly (e.g., artillery batteries) or into multifunction support brigades (e.g., aviation, sustainment, fires). These multifunction brigades frequently provide elements to the BCTs in a "plug-and-play" fashion. This ability to plug units into a BCT, in turn, makes BCTs better able to deploy and conduct a wide range of missions. However, this modularity also means that units that may have not historically been in a maneuver battalion's AO habitually could be today. For example, some of the elements (e.g., elements of the supply [A company] and maintenance [B company] that form an FSC historically would have been located with an FSB in the brigade support area. However, in a modular brigade, some elements of the former A and B companies are now in an FSC and the FSCs are located with the maneuver battalions, not in the brigade support areas.

Symmetrical Versus Asymmetrical Threats

The assignment policy for women was written at a time when most potential threats were believed to involve symmetrical TTPs. Here, symmetry implies that the enemy would have employed weapons and techniques that were similar to those that the U.S. military would use.[1] For example, the Soviets would have relied on long-range artillery, tanks, and mechanized infantry, and the U.S. military planned to use like weapons.[2] One result of this symmetrical warfare was that the armed forces were better able to predict where on the battlefield direct contact and, in turn, the highest probability of battle-related injuries

---

[1] At the time, there were differences in U.S. and Soviet doctrine, e.g., on the use of artillery during movement. However, at the most fundamental level, both doctrines were intended to fight using traditional, conventional tactics.

[2] An astute observer or planner of those potential large-scale armor battles would note that U.S. and NATO forces employed a different strategy in the development of arms, and so one may argue that these planned conflicts were not truly symmetrical. For example, the Soviets tended to develop less technologically sophisticated systems in mass quantities, and the United States designed and produced limited quantities of more sophisticated weapons. (A good example of this was the Soviet T-72 tank versus the U.S. M1A1 tank.) However, these battles would have fundamentally been conventional armor versus armor, not armor versus unconventional forces.

were likely to occur.[3] Thus, when the threat was symmetrical, commanders, strategists, and policy developers had a much clearer understanding of where direct contact was likely to occur and, thus, where on the battlefield soldiers and leaders would face the greatest responsibility to attack and destroy enemy forces.

Today, and most likely in the future, threats are likely to employ asymmetrical tactics against U.S. forces. Asymmetrical tactics are those designed by anti-U.S. forces to harm U.S. assets without going up against the "teeth" of U.S. defenses. For example, insurgents in Iraq have been more likely to target unarmored convoys or civilian locations, as opposed to our better-armed and -defended systems, such as the Abrams tank or Bradley fighting vehicle. These asymmetrical tactics inherently lead to situations in which direct contact is difficult to predict, i.e., direct contact may not happen only where infantry or armored forces are present, but anywhere in the theater of operations. In addition, on the asymmetrical battlefield, enemy direct-fire weapon systems may not cause the greatest degree of damage to U.S. forces. For example, Iraq insurgency forces have been much more likely to use IEDs (not a direct-contact weapon system) than to conduct direct engagement with direct-contact systems.

In the future, the line between symmetric conventional forces and asymmetric unconventional forces will most likely be blurred. It is possible that asymmetrical forces would be encountered only in stability and support operations similar to those in Iraq today and that other future battlefields may be more conventional in nature. However, several factors would suggest that this is unlikely and that future U.S. adversaries would employ at least a combination of symmetric and asymmetric tactics. First, most other land armies do not have the technological capabilities that could easily match U.S. forces. Consequently, a smart enemy would plan to employ techniques that would

---

[3]   Direct-contact weapon systems are those in which the operator (or shooter) can see and aim at the target while engaging the target. Examples of such direct-contact situations would include sniper attacks; a tank engaging another tank, vehicle, or foot soldiers; or standard rifleman engagement of threats. Indirect contact occurs when the shooter does not directly see the target. For example, artillery personnel do not directly see the target, and mines are indirect weapon systems because the placer of the mine does not directly engage the target.

harm U.S. interests and assets while bypassing the U.S. technological edge—that is, employ asymmetrical techniques. Second, the successes of U.S. land forces' training and equipment in both Operation Desert Storm and the first phase of Operation Iraqi Freedom against conventional forces were well published internationally. We should assume that potential adversaries are well aware of such successes and would plan tactics in an attempt to defeat such strengths. Third, there were accounts of unconventional approaches even during the early phases of Operation Iraqi Freedom (e.g., nonuniformed soldiers fighting against U.S. forces). Future opponents could escalate and improve on such techniques to more fully incorporate these asymmetric means into more symmetric wars. All of these factors suggest that future enemies will use a combination of tactics, and all parts of the theater will be dangerous and will require that leaders and soldiers are prepared for direct contact with the enemy.

Linear Versus Nonlinear Battlefields

The assignment policy was drafted at a time when battles were assumed to be linear. Battlefields today, and in the future, are assumed to be nonlinear. This linear versus nonlinear distinction has implications for the meaning and usefulness of the wording of the assignment policy. Most previous wars (including World War I, World War II, and Desert Storm) had linear battles and campaigns. At a basic level, in linear warfare, the advance of forces generally proceeded forward. This linear advance meant that there was a front (where direct contact occurred), two sides (or flanks) that were protected and generally were not where the bulk of the fighting occurred, and a rear area that was fairly secure against enemy land attack. As forces advanced, moving the front forward, they would clear and secure the land seized. Once the land was clear, the noncombat assets in the rear would move forward, establishing a new rear with protective defensives reestablished.

No rear was ever 100 percent secure from enemy attack; however, the rear could be characterized in two distinct and important ways from the front. First, the role of the forces in the rear was support-

or service-oriented. Their missions did not involve any planned direct combat with enemy forces. In addition, unlike those in the forward area of operations, the forces in the rear were in a fairly benign, rarely attacked area.

However, battlefields are now characterized as nonlinear with a 360-degree AO. BCTs, in particular, are no longer thought of as moving or advancing along a linear front, but instead are responsible for a more circular AO. In addition, with the Army's redesign of combat brigades into BCTs, the BCT would or could be responsible for a larger AO than ever before. These combined factors suggest that the AO on a nonlinear battlefield really has no well-defined "front" or "forward" area, and, consequently, there is no longer a defined rear in a brigade's AO, either. So, in today's and future wars, there is no safe rear AO, per se, where the vast majority of service and support units can be located. This situation does not mean that there are not some areas within an AO that are safer than others. For example, FOBs in Iraq could be considered some of the most secure places in a BCT's AO. However, FOBs are within the AO and are relative small areas when compared to the overall AO size. So, service and support forces may be secure in an FOB, similar to a rear area, but, once they leave the FOB, the level of security declines unlike that of the larger rear area of a linear battlefield.

The nonlinear, asymmetric nature of war means that there will be a much greater likelihood (than in previous wars) that forces that are not intended to engage in direct combat (such as maintenance or transportation units) will be confronted with lethal enemy actions.

Female Army Recipients of the Combat Action Badge

This appendix provides a descriptive account of the Army women who received the CAB as of August 2006.[1]

Enlisted CAB Recipients

Table H.1 lists the number of enlisted women, by pay grade, who received the CAB. Of the 1,521 female enlisted recipients, the vast majority was concentrated in the E-4, E-5, and E-6 pay grades, consistent with the distribution of enlisted women among those pay grades.[2]

As indicated in Table H.2, many female CAB recipients were military police, truck drivers, or in logistics and supply occupations. No other occupation accounted for more than 3 percent of the enlisted female recipients.

---

[1]   We are unable to distinguish from the data whether some of these women received the CAB for duty in Afghanistan.

[2]   Where women were listed twice in the data, we assumed that they were awarded the CAB once. Where women appeared in the list of enlisted personnel as well as in the officer or warrant officer lists, we counted them as enlisted awardees.

Table H.1
Enlisted Female CAB Recipients by Pay Grade

Pay Grade	Number	Percent Female Recipients	Percent Women at Pay Grade
E-1	8	0.5	0.8
E-2	7	0.5	5.6
E-3	56	3.7	13.8
E-4	464	30.5	31.9
E-5	580	38.1	22.3
E-6	260	17.1	14.4
E-7	107	7.0	8.7
E-8	25	1.6	2.0
E-9	14	0.9	0.6
Total	1,521	100	100

SOURCE: CAB data, Army Field Systems Division, Electronic Military Personnel Office (eMILPO), as of August 2006. End-strength data, Army G-1, as of the end of fiscal year 2006.

NOTE: Percentages do not sum to 100 due to rounding.

Table H.2
Enlisted Female CAB Recipients by Occupation

MOS	Occupation	Number	Percent Female Recipients
31B	Military police	159	10.3
42A	Human resource specialist	111	7.2
63B	Light-wheel vehicle mechanic	48	3.1
88M	Motor transport operator	183	11.9
91W (includes 68W)	Health care specialist	63	4.1
92A	Automated logistical specialist	158	10.3
92F	Petroleum supply specialist	53	3.5
92G	Food service operations	87	5.7
92Y	Unit supply specialist	119	7.8
All other codes		554	36.1
Total		1,535	100

SOURCE: Army Field Systems Division, eMILPO, as of August 2006.

Officer CAB Recipients

A total of 242 female officers received the CAB as of August 2006. Not surprisingly, the majority of female officer CAB recipients are in pay grades O-2, O-3, and O-4, consistent with the distribution of female officers. The officer recipients are listed by pay grade in Table H.3.

Officer recipients are listed by branch in Table H.4 and appear to be distributed relatively evenly across many of the officer branches.

Table H.3
Female Officer CAB Recipients by Pay Grade

Pay grade	Number	Percent Female Recipients	Percent Women at Pay Grade
O-1	1	0.4	9.6
O-2	41	16.9	14.8
O-3	129	53.3	41.0
O-4	43	17.8	18.8
O-5	23	9.5	11.2
O-6	5	2.1	4.5
Total	242	100	100

SOURCE: CAB data, Army Field Systems Division, eMILPO, as of August 2006. End-strength data, Army G-1, as of end of fiscal year 2006.

NOTE: Percentages do not sum to 100, both due to rounding and due to the small number of female officers in pay grades above O-6.

Table H.4
Female Officer CAB Recipients by Branch

Branch Code	Branch	Number	Percent Female Recipients
21	Engineers	20	8.3
25	Signal	23	9.5
31	Military police	20	8.3
35	Military intelligence	22	9.1
42	Adjutant general	9	3.7
66	Nurse	24	9.9
67	Medical service	14	5.8
74	Chemical	8	3.3
88	Transportation	21	8.7
91	Ordnance	27	11.2
92	Quartermaster	21	8.7
All other branches		33	13.6
Total		242	100

SOURCE: CAB data from Field Systems Division, eMILPO, as of August 2006.
NOTE: Percentages may not sum to 100 due to rounding.

Warrant Officer CAB Recipients

Twenty-five female warrant officers received the CAB as of August 2006. Only warrant officers in pay grades CW-2 to CW-4 received the CAB, and more than half of the recipients were CW-2s. The warrant officer recipients are listed by pay grade in Table H.5 and by occupation in Table H.6.

Table H.5
Female Warrant Officer CAB Recipients by Pay Grade

Pay Grade	Number	Percent Female Recipients	Percent Women at Pay Grade
WO-1	0	0	20.0
CW-2	14	56	40.1
CW-3	9	36	28.4
CW-4	2	8	8.5
CW-5	0	0	2.4
Total	25	100	100

SOURCE: CAB data, Army Field Systems Division, eMILPO, as of August 2006. End-strength data, Army G-1, as of end of FY 2006.

NOTE: Percentages may not sum to 100 due to rounding.

Table H.6
Female Warrant Officer CAB Recipients by MOS

MOS	Title	Number	Percent Female Recipients
152D	OH-58D scout pilot	6	24
152H	AH-64D attack pilot	1	4
153B	UH-1 pilot	1	4
153D	UH-60 pilot	2	8
154C	CH-47d pilot	1	4
155E	C-12 pilot	1	4
251A	Information system technician	2	8
351L	Counterintelligence technician	2	8
352N	Traffic analysis technician	1	4
915A	Unit maintenance technician	1	4
919A	Engineer equipment repair technician	1	4
920A	Property accounting technician	2	8
920B	Supply system technician	2	8
921A	Airdrop system technician	1	4
Unidentified	Unidentified	1	4
Total		25	100

SOURCE: CAB data, from Field Systems Division, eMILPO, as of August 2006.

Bibliography

Agostini, Luis R., "Women's Combat Support Role Could End in Iraq," *Marine Corps News*, May 19, 2005. As of October 25, 2006:
http://www.marines.mil/marinelink/mcn2000.nsf/lookupstoryref/2005519123339

Alderman, Marc I., *Women in Direct Combat: What Is the Price for Equality?* Ft. Leavenworth, Kan.: U.S. Army Command and General Staff College, School of Advanced Military Studies, December 1992.

Aspin, Les, Secretary of Defense, "Direct Combat Definition and Assignment Rule," memorandum, January 13, 1994.

"Bibliography of Women in the Military 1990–2003," Arlington, Va.: Women's Research and Education Institute, undated. As of October 24, 2006:
http://www.wrei.org/WomeninMilitary.htm

Blair, Anita K., "Gender-Integrated and Gender-Segregated Training," testimony before the House Armed Services Committee, Subcommittee on Military Personnel, Washington, D.C., March 17, 1999. As of February 11, 2007:
http://armedservices.house.gov/comdocs/testimony/106thcongress/99-03-17blair.htm

Brookside Associates, "Medical Support of Women in Military Environments," 2001. As of February 11, 2007:
http://www.brooksidepress.org/Products/OperationalMedicine/DATA/operationalmed/Manuals/enhanced/MedicalSupport.htm

Brosnan, James W., "House GOP Leaders Drop Women-in-Combat Provision," Scripps Howard News Service, May 25, 2005. As of October 25, 2006:
http://www.sitnews.us/0505news/052505/052505_shns_womencombat.html

Brower, J. Michael, "A Case for Women Warfighters," *Military Review*, November–December 2002, pp. 61–66. As of February 11, 2007:
http://usacac.leavenworth.army.mil/CAC/milreview/English/NovDec02/NovDec02/brower.pdf

Bucher, Merideth A., *The Impact of Pregnancy on U.S. Army Readiness*, Maxwell Air Force Base, Ala.: Air Command and Staff College, April 1999. As of February 11, 2007:
http://www.au.af.mil/au/awc/awcgate/acsc/99-016.pdf

Carter, Jeffrey A., "Females in the U.S. Army," Melbourne, Fla.: Florida Institute of Technology, November 1, 2001.

Center for Military Readiness, "Women in Land Combat," Livonia, Mich., report no. 16, April 2003a.

———, *Army Gender-Integrated Basic Training (GIBT): Summary of Relevant Findings and Recommendations, 1993–2002*, Livonia, Mich., May 2003b.

———, "Enlisted Women Opposed to Combat Assignments," Web page, September 3, 2003c. As of October 3, 2006:
http://www.cmrlink.org/WomenInCombat.asp?docID=204

———, "Hunter Admonishes Army on Women in Land Combat," Web page, June 1, 2005. As of September 27, 2006:
http://www.cmrlink.org/WomenInCombat.asp?DocID=249

———, "Army Still Violating Policy and Law on Women in Land Combat," *CMR Policy Analysis*, February 8, 2006a.

———, "Background and Facts: Women in or Near Land Combat," Web page, June 16, 2006b. As of October 23, 2006:
http://www.cmrlink.org/WomenInCombat.asp?docID=271

Cohen, Sharon, "Women Take on Major Battlefield Roles," Associated Press, December 3, 2006.

Congressional Research Service, *Persian Gulf War: Defense-Policy Implications for Congress*, Report 91-421 F, May 15, 1991. As of February 11, 2007:
http://digital.library.unt.edu/govdocs/crs//data/1991/upl-meta-crs-6963/91-421_1991May15.pdf

Cordesman, Anthony H., "The Lessons of the Iraq War: Main Report," Tenth Working Draft, Washington, D.C.: Center for Strategic and International Studies, July 2, 2003.

Council on Biblical Manhood and Womanhood, "Women in Combat: A Resolution from CBMW," undated Web page. As of October 3, 2006:
http://www.cbmw.org/resources/articles/combat.php

Culler, Kristen W., *The Decision to Allow Military Women into Combat Positions: A Study in Policy and Politics*, thesis, Monterey, Calif.: Naval Postgraduate School, June 2000.

Defense Department Advisory Committee on Women in the Services, "Forces Development and Utilization Subcommittee, Recommendation #1: Multiple Launch Rocket Systems (MLRS)," in *Defense Advisory Committee on Women in the Services 1999 Spring Conference Issue Book*, Washington, D.C., 1999. As of October 3, 2006:
http://www.dtic.mil/dacowits/issue_books/Spring99_IssueBk_toc.html

————, *Health Care Issues for Military Women and Female Military Family Members*, draft interim report, 2003.

Do, James J. W., *Understanding Attitudes on Gender and Training at the United States Air Force Academy*, thesis, Colorado Springs, Colo.: University of Colorado, March 30, 2006.

Donnelly, Elaine, "Women in Combat by Default?" *Washington Times*, May 6, 2006.

"Do We Really Need Mothers in Combat? Support Amendment to Uphold Ban on Women in Combat," *Eagle Forum*, May 13, 2005. As of September 27, 2006:
http://www.eagleforum.org/alert/2005/05-13-05.html

Feickert, Andrew, *U.S. Army's Modular Redesign: Issues for Congress*, Congressional Research Service, RL32476, updated May 5, 2006. As of October 25, 2006:
http://fpc.state.gov/documents/organization/67816.pdf

GlobalSecurity.org, "Operation Just Cause," Web page, last updated April 27, 2005a. As of February 11, 2007:
http://www.globalsecurity.org/military/ops/just_cause.htm

————, "Operation Urgent Fury," Web page, last updated April 27, 2005b. As of February 11, 2007:
http://www.globalsecurity.org/military/ops/urgent_fury.htm

Goldstein, Joshua S., *War and Gender: How Gender Shapes the War System and Vice Versa*, Cambridge, UK: Cambridge University Press, 2001. Excerpt from Chapter Two, as of October 3, 2006:
http://www.warandgender.com/wgwomcom.htm

Harris, Beverly C., Zita M. Simutis, and Melissa Meyer Gantz, *Women in the U.S. Army: An Annotated Bibliography*, Alexandria, Va.: U.S. Army Research Institute for the Behavioral and Social Science, May 2002.

Havron, Stephanie, "Women in Combat: Iraqi Freedom," combined bibliography, Maxwell Air Force Base, Ala.: Air University Library, May 2003. As of October 3, 2006:
http://www.au.af.mil/au/aul/bibs/iraq/wiraq.htm

Headquarters, U.S. Department of the Army, Army Regulation 600-13, Army Policy for the Assignment of Female Soldiers, March 27, 1992. As of February 11, 2007:
http://www.army.mil/usapa/epubs/pdf/r600_13.pdf

————, *Tactics, Techniques, and Procedures for the Forward Support Battalion (Digitized)*, FM 4-93.50, Washington, D.C., May 2, 2002.

————, *Heavy Brigade Combat Team Reconnaissance Squadron*, Washington, D.C., FM 3-20.96, March 15, 2005a.

————, Brigade Special Troops Battalion (HBCT) Table of Organization and Equipment Narrative, DOCNO 87305GFC12, effective October 16, 2005b.

————, Brigade Support Battalion (HBCT) Table of Organization and Equipment Narrative, DOCNO 63325GFC05, effective May 16, 2006a.

————, *The Brigade Combat Team*, FM 3-90.6, Washington, D.C., August 4, 2006b.

"House Drops Women in Combat Bill," Associated Press, May 26, 2005. As of September 27, 2006:
http://www.military.com/NewsContent/0,13319,FL_women_052605,00.html?ESRC=eb.nl

Iskander, Mona, "Female Troops in Iraq Redefine Combat Rules," *Women's eNews*, July 5, 2004. As of October 3, 2006:
http://www.womensenews.org/article.cfm/dyn/aid/1899

"Larsen Decries Proposal to Limit Women's Military Service," press release, Washington, D.C.: Office of Congressman Rick Larsen, May 18, 2005. As of September 27, 2006:
http://www.house.gov/list/press/wa02_larsen/pr_05182005_womeninservice.html

Lindstrom, Krista E., Tyler C. Smith, Timothy S. Wells, Linda Z. Wang, Besa Smith, Robert J. Reed, and Wendy E. Goldfinger, *The Mental Health of U.S. Military Women in Combat Support Positions*, report no. 04-29, San Diego, Calif.: Naval Health Research Center, August 25, 2004.

Maze, Rick, "House Panel Votes to Ban Women from Some Combat Support Jobs," *Army Times*, May 12, 2005. As of February 11, 2006:
http://www.armytimes.com/legacy/new/1-292925-843605.php

Moniz, Dave, "Female Amputees Make Clear That All Troops Are on the Front Lines," *USA Today*, April 28, 2005. As of February 11, 2007:
http://www.usatoday.com/news/nation/2005-04-28-female-amputees-combat_x.htm

Morden, Bettie J., *The Women's Army Corps, 1945–1978*, Washington, D.C.: U.S. Army Center of Military History, July 1989. As of October 3, 2006:
http://www.army.mil/cmh-pg/books/wac/index.htm

National Security–Foreign Relations Division, "TOPIC 2: Combat Role for Women in Flux," in *National Security–Foreign Relations Summary*, week ending April 22, 2005. As of October 25, 2006:
http://www.wilegion.org/programs/nsfr4-22-05.pdf

Office of the Assistant Secretary of Defense (Force Management Policy), "Gender of AC Enlisted Accessions," in *Population Representation in the Military Services, Fiscal Year 98*, Washington, D.C.: November 1999. As of October 3, 2006:
http://www.dod.mil/prhome/poprep98/html/2-gender.html

Office of the Deputy Chief of Staff for Personnel (Army G-1), Human Resources Policy, Women in the Army, undated Web page. As of October 3, 2006:
http://www.armyg1.army.mil/hr/wita.asp

Porter, Laurie M., and Rick V. Adside, *Women in Combat: Attitudes and Experiences of U.S. Military Officers and Enlisted Personnel*, thesis, Monterey, Calif.: Naval Postgraduate School, December 2001.

Presidential Commission on the Assignment of Women in the Armed Forces, *Women in Combat: Report to President*, Washington, D.C., November 15, 1992.

Public Law 109-163, National Defense Authorization Act for Fiscal Year 2006, January 6, 2006. As of February 11, 2007:
http://frwebgate.access.gpo.gov/cgi-bin/getdoc.cgi?dbname=109_cong_public_laws&docid=f:publ163.109.pdf

"Role of Women in the Theater of Operations," in *Conduct of the Persian Gulf War: Pursuant to Title V of the Persian Gulf Conflict Supplemental Authorization and Personnel Benefits Act of 1991 (Public Law 102-25)*, April 1992. As of October 24, 2006:
http://www.ndu.edu/library/epubs/cpgw.pdf

Sagawa, Shirley, and Nancy Duff Campbell, *Women in Combat*, Women in Combat Issue Paper, Washington, D.C.: National Women's Law Center, October 1992.

Scarborough, Rowan, "Hunter Bucks the Top Brass," *Washington Times*, May 16, 2005. As of October 3, 2006:
http://www.washtimes.com/national/20050516-124950-3391r.htm

———, "Parties Face Off Before Vote on Women in Combat," *Washington Times*, May 25, 2005. As of September 29, 2006:
http://washingtontimes.com/national/20050524-114246-3965r.htm

Schlesing, Amy, "Women's Jobs Open Up in War," *Arkansas Democrat-Gazette*, October 22, 2006. As of October 25, 2006:
http://www.nwanews.com/adg/News/170356

Schmitt, Eric, "First Woman in 6 Decades Gets the Army's Silver Star," *New York Times*, June 17, 2005, p. A16.

Segal, David R., and Mady Wechsler Segal, "America's Military Population," *Population Bulletin*, Vol. 59, No. 4, December 2004. As of February 11, 2007:
http://www.prb.org/Template.cfm?Section=Population_Bulletin1&template=/ContentManagement/ContentDisplay.cfm&ContentID=12460

"Senator Clinton Introduces Bill to Uphold Role of Women in Combat," press release, Washington, D.C.: Office of Senator Hillary Rodham Clinton, May 26, 2005. As of September 27, 2006:
http://www.senate.gov/~clinton/news/statements/details.cfm?id=239866&&

Sidoti, Liz, "House Committee Votes to Ban Women in Combat," *Capitol Hill Blue*, May 19, 2005a. As of September 27, 2006:
http://www.capitolhillblue.com/artman/publish/printer_6742.shtml

————, "House Panel Weighs Ban on Women in Combat," *SFGate.com*, May 19, 2005b. As of September 27, 2006:
http://www.sfgate.com/cgi-bin/article.cgi?f=/n/a/2005/05/18/national/w164259D50.DTL

Simons, Anna, "Women in Combat Units: It's Still a Bad Idea," *Parameters, U.S. Army War College Quarterly*, Summer 2001, pp. 89–100. As of October 3, 2006:
http://www.carlisle.army.mil/USAWC/parameters/01summer/simons.htm

Sullivan, Gordon R., U.S. Army Chief of Staff, and Togo D. West, Jr., Secretary of the Army, "Direct Combat Definition and Assignment Rule," memorandum to the Secretary of Defense, January 12, 1994.

The Technical Cooperation Program, "An Examination of Current Gender Integration Policies and Practices in TTCP Nations," TTCP/HUM/01/03, September 2001.

Triggs, Marcia, "Female Soldiers: Fighting, Dying for Their Country," Army News Service, March 17, 2004. As of February 11, 2007:
http://www.usarc.army.mil/63rsc/newsfemale.htm

Tyson, Ann Scott, "More Objections to Women-in-Combat Ban," *Washington Post*, May 18, 2005, p. A5. As of October 3, 2006:
http://www.washingtonpost.com/wp-dyn/content/article/2005/05/17/AR2005051701356.html

U.S. Army Women's Museum, Ft. Lee, Va., undated homepage. As of February 11, 2007:
http://www.awm.lee.army.mil/

U.S. Department of the Army, "2006 Posture Statement," February 10, 2006. As of February 11, 2007:
http://www.army.mil/aps/06

U.S. Department of Defense, *Department of Defense Dictionary of Military and Associated Terms*, Joint Publication 1-02, April 12, 2001 (as amended through January 5, 2007). As of February 11, 2007:
http://www.dtic.mil/doctrine/jel/new_pubs/jp1_02.pdf

U.S. General Accounting Office, *Basic Training: Services Are Using a Variety of Approaches to Gender Integration*, Washington, D.C., GAO/NSIAD-96-153, June 1996. As of February 11, 2007:
http://www.gao.gov/archive/1996/ns96153.pdf

———, *Gender Issues: Improved Guidance and Oversight Are Needed to Ensure Validity and Equity of Fitness Standards*, Washington, D.C., GAO/NSIAD-99-9, November 1998. As of February 11, 2007:
http://www.gao.gov/archive/1999/ns99009.pdf

———, *Gender Issues: Information on DoD's Assignment Policy and Direct Ground Combat Definition*, Washington, D.C., GAO/NSIAD-99-7, October 1998. As of February 11, 2007:
http://www.gao.gov/archive/1999/ns99007.pdf

———, *Gender Issues: Perceptions of Readiness in Selected Units*, Washington, D.C., GAO/NSIAD-99-120, May 1999. As of February 11, 2007:
http://www.gao.gov/archive/1999/ns99120.pdf

U.S. House of Representatives, House Armed Services Committee, press statement of Representative Duncan Hunter, chairman, "House Armed Services Committee Approves Fiscal Year 2006 Defense Authorization Bill," Washington, D.C., May 19, 2005a, p. 28.

———, "McHugh/Hunter Provision Limits Flexibility of Commanders During War," press release, May 24, 2005b. As of October 26, 2006:
http://wwwd.house.gov/hasc%5Fdemocrats/NDAA%20FY06%20info/WIC%205.24.05.pdf

———, "Hunter Statement on Department of Defense Direct Ground Combat Policy," press release, Washington, D.C., May 25, 2005c.

———, *Hearing on Recruiting and Retention and Military Personnel Policy, Benefits and Compensation Overview*, sample QFR format, Washington, D.C., April 6, 2006.

U.S. House of Representatives, House Armed Services Committee, Subcommittee on Military Personnel, *Statement and Status Report of the Congressional Commission on Military Training and Gender-Related Issues*, Washington, D.C., March 17, 1999.

U.S. Senate, S 1134-IS, To Express the Sense of Congress on Women in Combat, 109th Congress, 1st Session, May 26, 2005. As of February 11, 2007:
http://thomas.loc.gov/cgi-bin/query/z?c109:S.1134.IS:

Ursano, Robert J., Loree Sutton, Carol S. Fullerton, Ann E. Norwood, Sidney M. Blair, Michael P. Dinneen, M. Richard Fragala, Harry C. Holloway, James R. Rundell, Normund Wong, and James B. McCarroll, *Stress and Women's Health: Combat, Deployment, Contingency Operations and Trauma*, Bethesda, Md.: Uniformed Services University Health Sciences, MIPR no. 95MM5516, July 1996.

Webster's New Collegiate Dictionary, Springfield, Mass.: G & C Merriam Co., 1979.

West, Togo D., Jr., Secretary of the Army, "Increasing Opportunities for Women in the Army," memorandum to the Under Secretary of Defense, Personnel and Readiness, July 27, 1994.

Westrup, Darrah, "OIF/OEF Women," briefing, Ninth Annual VA Leadership Conference, Dallas, Texas, April 2006. As of October 25, 2006:
http://www.avapl.org/conference.html

Willens, Jake, "Women in the Military: Combat Roles Considered," Washington, D.C.: Center for Defense Information, August 7, 1996. As of October 3, 2006:
http://www.cdi.org/issues/women/combat.html

Wojack, Adam N., "Integrating Women into the Infantry," *Military Review*, November–December 2002, pp. 67–74. As of February 11, 2007:
http://usacac.army.mil/CAC/milreview/English/NovDec02/NovDec02/wojack.pdf

Women in Combat: An Annotated Bibliography, Ft. Leavenworth, Kans.: Command and General Staff College, Combined Arms Research Library, undated. As of October 3, 2006:
http://www-cgsc.army.mil/carl/resources/biblio/wic.asp

"Women in Combat: Lawmakers Draw New Line," MSNBC News Services, May 19, 2005. As of October 3, 2006:
http://www.msnbc.msn.com/id/7909442

"Women in the Armed Forces, Women in Combat," bibliography, Maxwell Air Force Base, Ala.: Air University Library. As of October 3, 2006:
http://www.au.af.mil/au/aul/bibs/women/womcom.htm

"Women Removed From Combat Support Roles," press release, Washington, D.C.: Office of Representative Vic Snyder, May 11, 2005. As of October 25, 2006:
http://www.house.gov/snyder/news_views/press_releases/pr05_11_05b.html

Wordsmyth English Dictionary-Thesaurus, Wordsmyth, updated continuously. As of February 11, 2007:
http://www.wordsmyth.net